THE SACRED AND
CIVIL CALENDAR OF THE
ATHENIAN YEAR

THE
SACRED AND CIVIL
CALENDAR OF THE
ATHENIAN YEAR

JON D. MIKALSON

PRINCETON UNIVERSITY PRESS

PRINCETON, NEW JERSEY

COPYRIGHT © 1975 BY PRINCETON UNIVERSITY PRESS
Published by Princeton University Press,
Princeton and London

All Rights Reserved

Library of Congress Cataloging in Publication data will
be found on the last printed page of this book

Publication of this book has been aided by
The Andrew W. Mellon Foundation

This book has been composed in Monotype Times Roman

Printed in the United States of America
by Princeton University Press,
Princeton, New Jersey

TO MARY

CONTENTS

PREFACE	ix
ABBREVIATIONS	xi
I. INTRODUCTION	
1. History of the Study	1
2. Meeting Days and Festival Days	3
II. PROLEGOMENA TO THE CALENDAR	
1. Denotation of Days	8
2. Validity of Restored Texts	10
3. Athenian Monthly Festival Days	13
III. CALENDAR OF THE ATHENIAN YEAR	25
IV. CONCLUSIONS	
1. Meeting Days of the Ekklesia	182
2. Festival Days and Meeting Days of the Ekklesia	186
3. Meeting Days of the Boule	193
4. Festival Days and Meeting Days of the Boule	196
5. Probable Dates for Major Athenian Festivals	197
APPENDIX I. CALENDAR OF DATED FINANCIAL TRANSACTIONS	207
APPENDIX II. PSEUDO-PSEPHISMATA	215

PREFACE

THE INVENTION and early use of calendars in ancient Greek society were an attempt to systematize and regularize the celebration of religious festivals within the city-state. The calendars, which differed markedly from one city-state to another, were thus religious institutions. These calendars were prescriptive in that they proclaimed that specific religious festivals and sacrifices were to be held every year on specific days of specific months. The original purpose of these calendars is reflected in the surviving fragments of the sacred calendars of the state of Athens and of the demes. But these calendars also provided a convenient framework by which one could record secular activities such as meetings and financial transactions. And thus in antiquity the Greeks used their calendars for both sacred and secular purposes: to prescribe the days of festivals, and to record the specific dates of secular activities.

It was my interest in Greek religion and in the Athenian cults and religious festivals which led me to the study of the Athenian calendar. In this I differ from other scholars who devote their attention to the Athenian calendar. They are, almost without exception, epigraphists whose primary interest is in restoring the fragmentary texts of dated decrees of the Athenian legislative assemblies.

This fundamental difference of approach to the Athenian calendar has resulted in two major differences in the treatment of the subject. Firstly, whereas previous studies have concentrated solely on the epigraphical and secular evidence, I have collected all the literary, scholiastic, lexicographic, and epigraphic evidence which relates to the dating of all sacred and secular activities in the Athenian year. Secondly, I have attempted to ascertain and gather only what can be proved conclusively concerning the dating of these festivals, meetings, and other activities. My goal has not been to increase our store of restored texts, but rather to determine and collect the evidence which can be proved valid as historical evidence. I have, therefore, proposed and followed rigid criteria for the acceptance of restored epigraphical texts as evidence. I have then used only that evidence which fulfills these criteria in discussing certain basic problems concerning the Athenian calendar. My criteria are conservative, and intentionally so. The great dissension which has marked recent calendric studies has been occasioned primarily by hypothetical theories based on

PREFACE

insufficient evidence and on hypothetically restored epigraphical texts. The only solution is to begin anew, and to sift and winnow the evidence in order to harvest that which is demonstrably valid. The result of my approach has been a collection of reliable evidence as to what religious and secular activities occurred on specific days of the Athenian year. This store of evidence has then made possible certain conclusions concerning the nature of the Athenian calendar.

Professor Sterling Dow has assisted me in this study from its inception to its conclusion. Professor Dow introduced me to the intricacies of Greek epigraphy, and with great interest directed my Ph.D. dissertation from which this study stems. Professor Zeph Stewart has encouraged and guided my studies in Greek religion. Professor Stewart and Professor Walter Burkert have kindly read earlier drafts of this study, and have offered countless valuable suggestions and criticisms. I am very much indebted to these scholars and mentors, as well as to Professors Walter R. Agard and Paul L. MacKendrick, whose influence as teachers has meant so much to me.

This study has been made possible by generous grants from Harvard University, The American School of Classical Studies at Athens, The University of Virginia, and The Center for Humanistic Sources at the University of Virginia. I should like to thank also Mrs. Joanna Hitchcock, the readers, and the staff of Princeton University Press who have undertaken and aided the publication of a text which provides so many technical problems of production.

And finally I wish to express my deepest gratitude to my wife, the *sine qua non* of my work, to whom this book is dedicated.

April 1, 1974   Jon D. Mikalson
Charlottesville, Virginia

ABBREVIATIONS

AJP	American Journal of Philology
BCH	Bulletin de Correspondance Hellénique
BSA	British School at Athens, Annual
Deubner, Feste	L. Deubner, Attische Feste, Berlin, 1932
Dinsmoor, Archons	W. B. Dinsmoor, The Archons of Athens in the Hellenistic Age, Cambridge, Massachusetts, 1931
Dittenberger, Sylloge³	W. Dittenberger, Sylloge Inscriptionum Graecarum³, Leipzig, 1915–1924
Dow, Prytaneis	S. Dow, Prytaneis, A Study of the Inscriptions Honoring the Athenian Councillors, Hesperia, Supplement I, 1937
Hesp	Hesperia, Journal of the American School of Classical Studies at Athens
HSCP	Harvard Studies in Classical Philology
HTR	Harvard Theological Review
IG	Inscriptiones Graecae
Inscriptions de Délos	P. Roussel and M. Launey, Inscriptions de Délos, Paris, 1937
Jacoby, FGrHist	F. Jacoby, Die Fragmente der griechischen Historiker, Berlin and Leiden, 1923–1958
Meritt, AFD	B. D. Meritt, Athenian Financial Documents of the Fifth Century, Ann Arbor, Michigan, 1932
Meritt, Year	B. D. Meritt, The Athenian Year, Berkeley and Los Angeles, California, 1961
Mommsen, Feste	A. Mommsen, Feste der Stadt Athen im Altertum, geordnet nach attischem Kalender, Leipzig, 1898
Pritchett, Choiseul	W. K. Pritchett, The Choiseul Marble, University of California Publications: Classical Studies, Vol. 5, 1970
Pritchett and Meritt, Chronology	W. K. Pritchett and B. D. Meritt, The Chronology of Hellenistic Athens, Cambridge, Massachusetts, 1940
Pritchett and Neugebauer, Calendars	W. K. Pritchett and O. Neugebauer, The Calendars of Athens, Cambridge, Massachusetts, 1947
RE	Pauly-Wissowa-Kroll, etc., Realencyclopädie der classischen Altertumswissenschaft
Roscher, Lex.	W. H. Roscher, Ausführliches Lexikon der griechischen und römischen Mythologie, Leipzig, 1884-1937
SEG	Supplementum Epigraphicum Graecum
TAPA	Transactions of the American Philological Association

THE SACRED AND
CIVIL CALENDAR OF THE
ATHENIAN YEAR

ἄλλα τ' εὖ δρᾶν φησιν, ὑμᾶς δ' οὐκ ἄγειν τὰς ἡμέρας
οὐδὲν ὀρθῶς, ἀλλ' ἄνω τε καὶ κάτω κυδοιδοπᾶν

Aristophanes *Nubes* 615–616

CHAPTER I
INTRODUCTION

1. HISTORY OF THE STUDY

During the course of this century the study of the Athenian calendar has become the subject of two distinct disciplines. The two disciplines have been practiced independently, and, although they are by nature closely related, the results of the study of one discipline have never been systematically applied to the other.

The two sections into which the study of the Athenian calendar has been divided may be designated as the "sacred" and the "civil." The sacred aspect has been studied primarily by historians of Greek religion in their attempts to date the Athenian religious festivals. Their sources are primarily literary or lexicographic. This study has, to date, culminated in the work of L. Deubner. In *Attische Feste* (1932) Deubner presented a calendar of all the Athenian festivals which could then be dated. The studies that have followed his monumental book have dealt with the dating of individual festivals.

The civil aspect has been studied primarily by epigraphists in their attempts to restore the preambles of Athenian decrees. Beginning in the second half of the fourth century B.C., the preambles of Athenian decrees specified the day and month of the meetings of the legislative assemblies responsible for the decrees. The texts of the preambles are often fragmentary, and interest in restoring the complete texts has led to specialized studies of the Athenian calendar by W. B. Dinsmoor, B. D. Meritt, W. K. Pritchett, and others.

In this century these two aspects of the Athenian calendar have been studied, for the most part, independently. When the historians of Greek religion, and especially Deubner, attempt to date a festival, they rarely take note of days on which a legislative assembly met. The epigraphists even less often take note of the festival days when they propose a restoration which would date a meeting of a legislative assembly.

A year calendar including all the known festivals and meetings is the single bridge which can span the gulf now existing between these two disciplines.

I. INTRODUCTION

August Mommsen, who laid the foundation for all later studies of the Athenian calendar and festivals, was the most recent scholar to study both aspects of the Athenian calendar. In 1864 (*Heortologie*, p. 93)[1] and in 1883 (*Chronologie*, pp. 143–156)[2] he published calendars of the type proposed *supra*, uniting into one single calendar of 360 days all the meetings and festivals then known. In 1898 he used this calendar effectively in his study of the Athenian festivals, *Feste der Stadt Athen*. Later studies and new data have outdated Mommsen's original calendar. In recent years certain details of the Athenian calendar have been studied intensively,[3] and several of Mommsen's citations must be revised. Secondly, Mommsen's own work on the dating of Athenian festivals was not completed until 1898, and this required further revision of his earlier calendars. Finally, and most importantly, the fund of epigraphical evidence has been greatly enriched since the publication of his calendars. The two major sources for this new epigraphical evidence are J. Kirchner's *Inscriptiones Graecae* II^2 and the publication of the valuable material discovered in the excavations of the Athenian Agora. Mommsen's calendars still remain valuable, however, as a collection of literary references to specific days.

The proposed new calendar of the Athenian year, including both meetings and festivals, will offer new data for a variety of historical questions. Any historical study which concentrates on exact chronology (e.g., of the political maneuverings which led to the Peace of Philokrates) obviously requires such a calendar. Historical questions such as these, however, lie beyond the compass of the present study.

Certain basic questions concerning the nature of the Athenian calendar arise, and for the first time can be examined. From the proposed civil calendar of meetings one can determine if the Athenian legislative assemblies could meet on any day of the year (as the restorations of some epigraphists would suggest), if they could meet only on certain days, or if the days of meetings tended to follow a

---

[1] A. Mommsen, *Heortologie, Antiquarische Untersuchungen über die städtischen Feste der Athener*, Leipzig, 1864 (reprinted Amsterdam, 1968).

[2] A. Mommsen, *Chronologie, Untersuchungen über das Kalenderwesen der Griechen, insonderheit der Athener*, Leipzig, 1883.

[3] The most useful and intelligible study of the Athenian calendar is still *The Calendars of Athens* by W. K. Pritchett and O. Neugebauer.

general pattern. These questions should be of particular interest to those who propose restorations which date meetings of legislative assemblies.

When the civil and sacred calendars are juxtaposed, the nature of their basic relationship can be systematically studied for the first time. Was the Athenian calendar similar to the Roman calendar, in that meetings could not occur on festival days? And, finally, the combined civil and sacred calendar will provide a new approach to several unanswered questions concerning monthly festivals and the proper dating of certain annual festivals.

2. Meeting Days and Festival Days

The most basic question, and that upon which the answers to many other questions depend, can be simply stated: Were meetings of legislative assemblies held on festival days? For example, would a meeting of the Ekklesia be held on Anthesterion 12, a day of the Anthesteria?

Common sense would, I believe, give a negative answer to this question. It is a nearly universal custom that certain secular activities such as legislative assemblies are suspended on the days of major religious celebrations, especially in societies which have a homogeneous religion. In the Roman calendar this procedure was specified in detail.

One might expect that if a matter of sufficient urgency arose, a meeting of a legislative assembly might be held despite the religious celebration. But the Athenian calendar was provided with a mechanism which enabled the Athenians to avoid such a conflict. The archon had the authority to intercalate one or more days at any time during the year. If, for example, on Posideon 25 a matter of great urgency arose which required a meeting of the Ekklesia on the following day, the archon could avoid conflict with the Haloa of Posideon 26 by intercalating a day between Posideon 25 and 26. The meeting of the Ekklesia could then be held on the intercalated day, and on the following day (Posideon 26) the Haloa could be celebrated. And also, the vast majority of the dated Athenian decrees which survive are decrees honoring individuals, and they therefore do not concern a matter of such urgency as to necessitate

I. INTRODUCTION

a meeting on a festival day. Thus, because of the process of intercalation and because of the nature of the decrees, we would not expect meetings to be attested for festival days.

When we turn from these rather general considerations to the sources, we find that the orators give a clear indication that it was considered irregular for a meeting of the Ekklesia to be held on a festival day.

Aeschines 3.66–67:

Δημοσθένης... γράφει ψήφισμα... ἐκκλησίαν ποιεῖν τοὺς πρυτάνεις τῇ ὀγδόῃ ἱσταμένου τοῦ Ἐλαφηβολιῶνος μηνός, ὅτ᾽ ἦν τῷ Ἀσκληπιῷ ἡ θυσία καὶ ὁ προαγών, ἐν τῇ ἱερᾷ ἡμέρᾳ, ὃ πρότερον οὐδεὶς μέμνηται γεγονός, τίνα πρόφασιν ποιησάμενος; "Ἵνα," φησίν, "ἐὰν παρῶσιν ἤδη οἱ Φιλίππου πρέσβεις, βουλεύσηται ὁ δῆμος ὡς τάχιστα περὶ τῶν πρὸς Φίλιππον."

Demosthenes 24.26:

[Τιμοκράτης] οὔτε γὰρ ἐξέθηκε τὸν νόμον, οὔτ᾽ ἔδωκεν εἴ τις ἐβούλετ᾽ ἀναγνοὺς ἀντειπεῖν, οὔτ᾽ ἀνέμεινεν οὐδένα τῶν τεταγμένων χρόνων ἐν τοῖς νόμοις, ἀλλὰ τῆς ἐκκλησίας, ἐν ᾗ τοὺς νόμους ἐπεχειροτονήσατε, οὔσης ἑνδεκάτῃ τοῦ Ἑκατομβαιῶνος μηνός, δωδεκάτῃ τὸν νόμον εἰσήνεγκεν, εὐθὺς τῇ ὑστεραίᾳ, καὶ ταῦτ᾽ ὄντων Κρονίων καὶ διὰ ταῦτ᾽ ἀφειμένης τῆς βουλῆς, διαπραξάμενος μετὰ τῶν ὑμῖν ἐπιβουλευόντων καθέζεσθαι νομοθέτας διὰ ψηφίσματος ἐπὶ τῇ τῶν Παναθηναίων προφάσει.

The complaint of both orators is essentially the same; their opponent in an underhanded manner had arranged for a meeting of the Ekklesia on a festival day. Even if we allow for rhetorical exaggeration, the implication that it was irregular for the Ekklesia to meet on these festival days is unmistakable.

Evidently no civil or religious law prohibited meetings on a festival day. If there had been such a law, these accomplished orators would not have overlooked the opportunity to charge their opponents with παρανομία or ἀσέβεια. Aeschines suggests that it was a violation of convention and tradition, not of law. This is an important distinction, but in matters of religion the force of tradition

MEETING DAYS AND FESTIVAL DAYS

was very strong, and in many cases lacked only the punitive force of law.

The dated decrees preserved on stone are usually honorary, and they would not, for the most part, be subject to the political contrivances which Demosthenes and Aeschines are describing. If, therefore, Demosthenes and Aeschines complain of the unusual procedure of a meeting of the Ekklesia on a festival day for such measures, it is highly improbable that the honorary decrees preserved on stone were passed in meetings held on festival days.

Modern scholars, however, assume that meetings of legislative assemblies frequently occurred on festival days. This is an implicit assumption of some epigraphists, especially B. D. Meritt, who restore preambles of decrees to give meetings on festival days. W. S. Ferguson (*Hesp* 1948, p. 134) states the point explicitly: "In 1898 Julius Dutoit...tabulated the known instances of conflict in Athens between meetings of the ecclesia and the occurrences of religious festivals. They were frequent: a sacred day, ἱερὰ ἡμέρα, in Athens was not ἀποφράς, nefastus."

Ferguson bases his statement on the study of Julius Dutoit, *Zur Festordnung der grossen Dionysien*,[4] the only published study dealing with the relationship of meetings to festival days. Dutoit wished to demonstrate that Elaphebolion 14 may have been a day of the City Dionysia, despite the fact that at least once the Ekklesia met on that day (Thucydides 4.118). In order to demonstrate that a meeting of the Ekklesia did not exclude the possibility that the day was a festival day, he enumerated all the cases in which a decree was passed on a festival day. This study alone has provided the basis of the assumption that meetings could occur on festival days.

Dutoit (pp. 38–39) tabulated the instances of conflict in a chart, which I have reproduced below. I have updated his references so as to give the corresponding reference to *Inscriptiones Graecae* II^2. I have indicated by question marks information which, in terms of present evidence, is either incorrect or hypothetical. For detailed discussions of the evidence, see the specific days in the calendar. A question mark in the "Feste" column indicates that the festival is either misdated or hypothetically dated. A question mark in the "Belegstelle" column indicates that the date in the inscription has been restored. It is obviously crucial that, in presenting instances in

[4] J. Dutoit, *Zur Festordnung der grossen Dionysien*, Speyer, 1898.

I. INTRODUCTION

which decrees were passed on festival days, one must be absolutely certain of the date of both the festival and the decree. If one or the other is not established beyond all reasonable doubt, then nothing is proved.

Monat	Tag	Das auf diesen Tag fallende Fest	Belegstelle
Boedromion	6	Marathonien	*IG* II^2 1039
,, ,,	18	Eleusinien (?)	*IG* II^2 657
,, ,,	18	,, ,, (?)	*IG* II^2 787 (?)
,, ,,	20	,, ,,	*IG* II^2 768 (?)
Pyanopsion	6	Kybernesien (?)	*IG* II^2 1014
,, ,,	11	Thesmophorien	*IG* II^2 1006
,, ,,	29	Apaturien (?)	*IG* II^2 353 (?)
Maimakterion	21	Fest des Zeus (?) Georgos	Diog. Laert. 7.1.10
Posideon	9	Piraeen (?)	*IG* II^2 1009
,, ,,	11	,, (?)	*IG* II^2 666
Gamelion	8	Lenaeen (?)	*IG* II^2 1012
,, ,,	11	,, (?)	*IG* II^2 450
,, ,,	11	,, (?)	*IG* II^2 886 (?)
,, ,,	11	,, (?)	*IG* II^2 968 (?)
,, ,,	11	,, (?)	*IG* II^2 1011
Anthesterion	19	Kleine Mysterien (?)	*IG* II^2 978
Elaphebolion	5	Gr. Dionysien (?)	*IG* II^2 646 (?)
,, ,,	8	Asklepieia und Proagon	Aeschines 3.67
,, ,,	9	Gr. Dionysien (?)	*IG* II^2 1008
,, ,,	14	,, ,,	Thuc. 4.118
Mounichion	16	Mounichien	*IG* II^2 644
,, ,,	16	,, ,,	*IG* II^2 905 (?)
,, ,,	19	Olympieen	*IG* II^2 775
Thargelion	25	Plynterien	*IG* II^2 485 (?)

Dutoit claims that twenty of the two hundred decrees he surveyed (i.e., 10 percent) were passed on festival days. This is the basis of Ferguson's statement that such conflicts were "frequent." The question marks in the chart reveal, however, how much of Dutoit's evidence has proved to be either incorrect or hypothetical. Of his twenty alleged instances of conflict, only six have proved to be valid, thus reducing the percentage of conflicts to 3. But even these numerical statistics are unsatisfactory, because at least one instance (Elaphebolion 8) is admitted by the source (Aeschines 3.67, cited *supra*) to be an exceptional situation. In another instance (Elaphebolion 14) Dutoit used his hypothesis to confirm the hypothesis itself.

This brief re-examination of Dutoit's study suggests that in fact meetings on festival days were very infrequent (3 percent), the opposite of what Dutoit was attempting to demonstrate. Fortunately, we have far more evidence than was available to Dutoit in 1898, and utilizing this new evidence we shall examine in the calendar *infra* the relationship of meetings and festival days.

All the evidence considered thus far indicates that meetings on festival days were exceptional and infrequent. A final conclusion will depend, of course, on the study of the calendar presented *infra*, but I have taken the liberty of presenting preliminary conclusions about festivals in the summary of each month, basing these conclusions on the assumption that meetings did not commonly occur on festival days. A final statement will be made on these preliminary conclusions only after we have resolved the basic questions concerning the occurrence of meetings on festival days.

CHAPTER II

PROLEGOMENA TO THE CALENDAR

BEFORE THE presentation of the annual calendar certain topics, some of which have aroused controversy, must be discussed. Two disputed points which are of importance for the present study are the validity of certain epigraphical restorations and the proper denotation of certain days. Also the Athenian monthly festivals, which have never been studied as a distinct group, must be surveyed before they are incorporated, as they must be, into the Athenian annual calendar. Certain comments relative to the format of the annual calendar will also be included among these *prolegomena*.

1. DENOTATION OF DAYS

Throughout this study the days of the Athenian month will be designated with arabic numerals corresponding to the following Greek designations:

1 - νουμηνία
2 - δευτέρα ἱσταμένου
3 - τρίτη ἱσταμένου
4 - τετράς ἱσταμένου
5 - πέμπτη ἱσταμένου
6 - ἕκτη ἱσταμένου
7 - ἑβδόμη ἱσταμένου
8 - ὀγδόη ἱσταμένου
9 - ἐνάτη ἱσταμένου
10 - δεκάτη ἱσταμένου
11 - ἑνδεκάτη
12 - δωδεκάτη
13 - τρίτη ἐπὶ δέκα
14 - τετράς ἐπὶ δέκα
15 - πέμπτη ἐπὶ δέκα
16 - ἕκτη ἐπὶ δέκα
17 - ἑβδόμη ἐπὶ δέκα
18 - ὀγδόη ἐπὶ δέκα
19 - ἐνάτη ἐπὶ δέκα

20 - εἰκοστή, δεκάτη προτέρα
21 - δεκάτη ὑστέρα
22 - ἐνάτη φθίνοντος,
    ἐνάτη μετ᾽ εἰκάδας
23 - ὀγδόη φθίνοντος,
    ὀγδόη μετ᾽ εἰκάδας
24 - ἑβδόμη φθίνοντος,
    ἑβδόμη μετ᾽ εἰκάδας
25 - ἕκτη φθίνοντος,
    ἕκτη μετ᾽ εἰκάδας
26 - πέμπτη φθίνοντος,
    πέμπτη μετ᾽ εἰκάδας
27 - τετράς φθίνοντος,
    τετράς μετ᾽ εἰκάδας
28 - τρίτη φθίνοντος,
    τρίτη μετ᾽ εἰκάδας
29 - δευτέρα φθίνοντος,
    δευτέρα μετ᾽ εἰκάδας
30 - ἕνη καὶ νέα

DENOTATION OF DAYS

The days of each month are numbered as though a month contained thirty days. This is a somewhat artificial systematization, because an Athenian month might have either twenty-nine or thirty days, depending on the first observance of the new moon.[1] If the month had twenty-nine days, then the twenty-ninth day was the ἕνη καὶ νέα. If the month had thirty days, then the thirtieth day was the ἕνη καὶ νέα.

The system of numerical notation *supra* will not, however, either deceive the reader or falsify any of the results. In the first place, whether a month was hollow (having twenty-nine days) or full (having thirty days) does not affect the numbering of days νουμηνία through τρίτη φθίνοντος, and thus there is no question concerning the notations for days 1–28. Also, δευτέρα φθίνοντος, when it occurs, must always be the twenty-ninth day. The only artificiality is to designate ἕνη καὶ νέα always as 30. It is necessary to do this in order to distinguish ἕνη καὶ νέα from δευτέρα φθίνοντος. It is also in perfect accord with the Athenian practice, whereby if the month were hollow, δευτέρα φθίνοντος and not ἕνη καὶ νέα was omitted.[2] So too in this study the last day in a hollow month (the ἕνη καὶ νέα) will be designated as 30, and day 29 will be considered as the day omitted.

The second major question involving the denotation of days concerns the dates including the phrase μετ᾽ εἰκάδας. For years the question was whether the calculation should be forward (τρίτη μετ᾽ εἰκάδας denoting 23) or backward from the end of the month (τρίτη μετ᾽ εἰκάδας denoting 28). Meritt in 1935 (*Hesp* 1935, pp. 525–561) attempted to prove formally that although the μετ᾽ εἰκάδας dates were usually calculated backward, there were some years in which forward calculation was used. In 1947 Pritchett and Neugebauer (*Calendars*, pp. 23–30) convincingly demonstrated that the evidence would allow retrograde calculation of μετ᾽ εἰκάδας dates in every year. In 1961 Meritt still defended his original proposition (*Year*, pp. 38–59). But in 1964 Meritt finally abandoned belief in forward calculation.[3] I, therefore, have confidently

---

[1] The precise method for determining the length of a month is disputed: see Pritchett and Neugebauer, *Calendars*, pp. 11–14, and Meritt, *Year*, pp. 16–37.

[2] Pritchett and Neugebauer, *Calendars*, p. 31. See also Pritchett, *BCH* 88 (1964), pp. 463–467. For the claim that it was not the twenty-ninth day which was omitted in a hollow month, but either the twenty-first or twenty-second, see Meritt, *AJP* 95 (1974), pp. 268–279.

[3] *TAPA* 95 (1964), p. 256, note 200.

II. PROLEGOMENA TO CALENDAR

employed retrograde calculation of μετ' εἰκάδας dates throughout this study.[4]

2. VALIDITY OF RESTORED TEXTS

The restoration of epigraphical texts has become a major factor in the study of the Athenian calendar. Hundreds of preambles to decrees are fragmentary, and there are all degrees of incompleteness in what survives. The restorations, especially of the calendric information, are often made on the basis of very slender evidence or of preconceived theories concerning the nature of the calendar itself. As a result, the correctness of individual restorations is often disputed. Because of the sparseness of evidence for some restorations, it is not uncommon for one scholar over a period of years to propose two or even three different restorations for the same inscription. The value as historical evidence of most of these restored texts is negligible. Some of the restorations may be, and probably are, correct, but in many cases there is no evidence to distinguish a correct from an incorrect restoration. A restoration may be rejected by the author of the restoration himself, but even his rejection is no guarantee that the restoration is incorrect; later he may readopt this same rejected restoration.[5]

In order to liberate my study from the uncertainty and constant revision which characterize some studies of the Athenian calendar, I have accepted as evidence only those restorations which are demonstrably correct. This conservative approach to the evidence is intended to provide a foundation of historical fact as distinguished from unproved theory, two elements which have often been confused in the study of the Athenian calendar.

I have accepted as evidence only restored texts in which the preserved formulae, the preserved prytany number, or, less frequently, the space available guarantees that the restoration of the month and day is correct. In many cases a combination of these elements establishes the validity of the restoration.

---

[4] The best discussion of the problem is by Pritchett and Neugebauer, *Calendars*, pp. 23–30.

[5] E.g., Meritt, *Hesp* 1944, pp. 234–241, no. 6; *Year*, pp. 119–120; *Hesp* 1963, pp. 425–432.

VALIDITY OF RESTORED TEXTS

Hesp 1961, pp. 289-292, no. 84, lines 3-7, presents a clear example of restoration on the basis of preserved formulae.

['Επὶ 'Αρχίππ]ου ἄρχοντος ἐπὶ τῆς Λ[ε]-
[ωντίδος πέμπτης π]ρυτανείας ἧι Σωκρατ[.]
[......¹⁴......] δης ἐγραμμάτευεν· Μαι[μ]-
[ακτηριῶνος ἕνει] καὶ νέαι, πέμπτει τῆς[πρ]-
[υτανείας· ἐκκλησ]ία

The preambles of Athenian decrees consist of a regular series of fixed items, and for certain periods one can expect the designation of the month and day after the designation of the secretary. In lines 5-6 MAI [- - -] can only be restored as the month Maimakterion. The preserved letters allow no other month. The designation of the days was also standardized (see Denotation of Days *supra*), and the preserved letters KAINEAI allow only the restoration [ἕνει] καὶ νέαι. Thus the completion of the formulae guarantees that this restoration to give Maimakterion 30 is correct.

IG II² 796, lines 1-4, provides an example of a month restored on the basis of the partially preserved prytany number.

['Επ' Εὐξενίππου ἄρχοντος ἐπὶ] τῆς 'Ερεχθεῖδος δευ[τ]-
[έρας πρυτανείας ἧι ...⁷... ο]ς Λύκου 'Αλωπεκῆθε[ν]
[ἐγραμμάτευεν· Μεταγειτνιῶνο]ς δεκάτει ὑστέραι,
[μιᾶι καὶ εἰκοστῆι τῆς πρυτανεί]ας· ἐκκλησία

The year of the decree, 305/4 B.C., can be demonstrated to have been ordinary,[6] and therefore the second prytany must have encompassed days ca. 29-60 of the year. The day of the month (21) is preserved in line 3, and only Metageitnion 21 would fall in days 29-60 of the year. Boedromion 21 would be day 80 ± 1 of the year, and Hekatombaion 21 would be day 21 of the year. Thus the restoration to give Metageitnion 21 is established.

Restorations based on the amount of space available are often the subject of much debate. Conflicting claims of measurement, size and spacing of letters, uninscribed spaces and scribal errors are often introduced. For this reason I have been very conservative in accepting restorations based primarily on the amount of space available. In general, relatively short spaces with established termini in

---
[6] Pritchett and Neugebauer, *Calendars*, p. 79.

II. PROLEGOMENA TO CALENDAR

stoichedon inscriptions offer the most security. *IG* II^2 769, lines 1-6, presents a clear example of this.

['Επ' Άντιμάχου] ἄρχοντος ἐ[πὶ τῆ]ς Αἰ[αν]-
[τίδος τετά]ρτης πρυτανεί[ας ἧι] Χαι[ρι]-
[γένης Χαι]ριγένου Μυρρι[νούσι]ος ἐ[γρ]-
[αμμάτευεν·] Πυανοψιῶνος [ἕκτ]ει μετ' εἰ-
[κάδας, πέμπ]τει καὶ εἰκοστεῖ τῆς πρυτ-
[ανείας· ἐκκ]λησία κυρία

In line 4 [ἕκτ]ει must be restored, because it is the only numeral with the requisite number of letters.

Using the criteria detailed *supra*, either separately or in combination, I have rejected or accepted restored texts as historical evidence. Those rejected are listed for the purpose of reference under the heading "Restoration to give (month-day)."

I have not presented the specific reasons for acceptance or rejection of each of the inscriptions considered. Protracted discussion of each one would be repetitious and unnecessary. The restorations rejected have been rejected because they cannot be proved to be correct by the criteria specified *supra*. Should the reader have reservations concerning a particular inscription, he may consult the source cited under "Restorations to give..." and apply the criteria himself. Only in particularly complex cases have the reasons for accepting a given restoration been presented. A reader with an understanding of the criteria specified *supra* will immediately recognize the basis of the validity of each of the restorations accepted.

In the calendar sufficient amount of the text of each inscription is presented to provide (where practical) the following information:

1. The extent of the restoration. The number of letters which must be restored is a factor in evaluating a restoration. It may be stated as a general rule (with many exceptions) that the longer the restoration, the less certain it is.

2. The year date of the inscription. In many cases the year date of the inscription is determined by the archon or the secretary. Often these have been restored, and thus the year date may be only as certain as the restoration. I have not evaluated each restoration as to year date, but have presented the generally accepted restoration. I have for the most part followed the archon list presented by Meritt, *Year*, pp. 231-238.

MONTHLY FESTIVAL DAYS

3. The number and day of the prytany. These are often restored, and are often uncertain. Because of the Athenians' frequent use of intercalation, neither the prytany date nor the day date can be restored simply on the basis of one another. I have not attempted to evaluate or revise the restorations of prytany numbers and days, but have only provided the text from the source cited.

4. The nature of the meeting, viz. whether it was a meeting of the Boule, Ekklesia, or another group. Also whether it was a meeting of the Ekklesia Kyria. Finally, the place of the meeting. This information is also subject to the uncertainties of restoration, and I have provided the text of the source cited.

I have not attempted to present a complete bibliography for each inscription considered. For the most part, only the original source of each restoration has been cited. The history of each restoration has not been traced beyond Kirchner's *Inscriptiones Graecae* II[2]. When Kirchner himself was not responsible for the restoration, he cited the source from which he obtained it.

3. ATHENIAN MONTHLY FESTIVAL DAYS

Among the Greeks certain days of each month were sacred. For example, in much of the Greek world the seventh day of each month was sacred to Apollo. These monthly festival days obviously form an integral part of the Athenian calendar, and must be surveyed before the presentation of the annual Athenian calendar. W. Schmidt (*Geburtstag im Altertum*)[7] and M. P. Nilsson (*Die Entstehung und religiöse Bedeutung des griechischen Kalenders*[2], pp. 35–44)[8] have collected much of the evidence for these monthly festivals, and have provided the basis for this study.

In the following survey of Athenian monthly festival days, I differ from previous studies of the monthly festivals on three basic points.

Firstly, I reject the "decadal" theory which claims that a day maintains its character in each decade, e.g., that if the fifth day is of a certain character, then also the fifteenth and the twenty-fifth days will be of the same character. This theory, which was suggested by

---

[7] W. Schmidt, *Geburtstag im Altertum, Religionsgeschichtliche Versuche und Vorarbeiten*, Vol. 7, Giessen, 1908.

[8] M. P. Nilsson, *Die Entstehung und religiöse Bedeutung des griechischen Kalenders*[2], Lund, 1962.

II. PROLEGOMENA TO CALENDAR

Hesiod *Op.* 782–789, 794–801, and 810–813, and which was used by Mommsen throughout his work, has no support in the attested Attic religious calendars or festivals. The lines of Hesiod which provided the germ for this theory may only be a reflection of archaic numerology.[9]

Secondly, monthly festival days which were celebrated only by private religious associations have been omitted from this study. Several private associations met once each month for a sacrifice and a banquet, and the members received their names from the day on which they met, e.g., Noumeniastai, Dekadistai, and Eikadistai.[10] Meetings of these and of other private associations, without further evidence, do not indicate monthly festival days of the state.

Thirdly, and most importantly, I have used only evidence which is directly linked with Athens or which can be linked with Athens by other evidence. The specific days which were sacred and the deities to whom they were consecrated could differ, like most beliefs of Greek religion, from city to city. It is thus important in establishing the Athenian calendar that we be able to provide evidence that the day was sacred *in Athens*. A scholiast, when he states, e.g., that the seventh day was sacred to Apollo, may or may not have been referring to Athens. In order to use such statements in a study of the Athenian calendar, we must be able to provide evidence that the day was in fact sacred in Athens. The result of limiting the evidence to that which refers to or can be associated with Athens is that several references, especially by non-Athenians such as Hesiod and Dionysios of Halikarnassos, have been omitted. In each case of omission there is no evidence that the source is providing information which refers to the Athenian calendar.

Day 1

Throughout the Greek world the first day of every month was entitled the Noumenia,[11] the "new-moon-day." Plutarch (*Mor.* 828A) designates it as the "holiest of days."

The festival in Athens was characterized by a large market (Aristophanes *Eq.* 43–44 and *Vesp.* 169–171) and by the pleasures of

---

[9] A. E. Samuel, *TAPA* 97 (1966), pp. 425–426.
[10] F. Poland, *Geschichte des griechischen Vereinswesens*, pp. 64 and 252–253.
[11] See discussion of Noumenia by author in *HTR* 65 (1972), pp. 291–297.

the palaestra (Aristophanes *Ach.* 999) and of the banquet (Athenaios 12.551F). A specific feature of the private religious practices of this day appears to have been the placing of frankincense on the statues of the gods (Aristophanes *Vesp.* 94–96 and the scholion to this passage). Demosthenes (25.99) indicates that the citizens went to the Acropolis on this day, and in all probability this was for the state presentation of the Epimenia[12] as described by Herodotos (8.41).

It is noteworthy that no annual religious festival is attested to have occurred on the Noumenia or to have included it. Evidently the Athenians not only considered the day as a holy day, but also did not allow other religious festivals to infringe upon it.

Day 2

The second day of each month was considered the day of the ἀγαθὸς δαίμων.

Hesychios ἀγαθοῦ δαίμονος πόμα·
  τὸ μετὰ τὸ δεῖπνον ἄκρατον πινόμενον παρὰ 'Αθηναίοις·
  καὶ τὴν β ἡμέραν οὕτως ἐκάλουν.
Suda ἀγαθοῦ δαίμονος·
  καὶ ἡμέραν δὲ τὴν δευτέραν τοῦ μηνὸς οὕτως ἐκάλουν.

Rohde (*Psyche*[8], English tr. by W. B. Hillis, pp. 207–208, note 133) views the ἀγαθὸς δαίμων as a chthonic figure, and is supported in this by Plutarch *Mor.* 270A, καὶ γὰρ Ἕλληνες, ἐν τῇ νουμηνίᾳ τοὺς θεοὺς σεβόμενοι, τὴν δευτέραν ἥρωσι καὶ δαίμοσιν ἀποδεδώκασι.... Roscher (*Lex.* I, Columns 98–99) follows Athenaios 15.675B in associating the ἀγαθὸς δαίμων with Dionysos: καὶ διὰ τοῦθ' οἱ Ἕλληνες τῷ μὲν παρὰ δεῖπνον ἀκράτῳ προσδιδομένῳ τὸν ἀγαθὸν ἐπιφωνοῦσι δαίμονα, τιμῶντες τὸν εὑρόντα δαίμονα· ἦν δ' οὗτος ὁ Διόνυσος. There is, in fact, a sacrifice to Dionysos on this day in Anthesterion in the Erkhia calendar (see Anthesterion 2).

The scholiast to Aristophanes *Nub.* 616 designates Day 2 as a day of honoring Poseidon: καὶ γὰρ ἐν τῇ πρώτῃ ἡμέρᾳ τεταγμένον ἦν τὸν Δία τιμᾶν, ἐν δὲ τῇ δευτέρᾳ τὸν Ποσειδῶνα, καὶ τοὺς ἄλλους θεοὺς καθεξῆς. This scholion must be incorrect, because the eighth day of the month was Poseidon's day (see Day 8). The scholiast's statement

---

[12] Suda ἐπιμήνια and Hesychios ἐπιμήνιοι.

II. PROLEGOMENA TO CALENDAR

concerning the Noumenia, and his statement καὶ τοὺς ἄλλους θεοὺς καθεξῆς attempt to introduce a systematization which is lacking in the Athenian religious calendar.

Day 3

Harpokration (also *Etym. Magn.* 767.40 ff. and Suda) states that this day was Athena's birthday: τριτομηνίς· Λυκοῦργος ἐν τῷ Περὶ τῆς ἱερείας· τὴν τρίτην τοῦ μηνὸς τριτομηνίδα ἐκάλουν· δοκεῖ δὲ γενέθλιος τῆς 'Αθηνᾶς. In Bekker's *Anecd.* 1.306.32 an abbreviated citation indicates that there was a public celebration on this day: τριτομηνίς· ἑορτὴ ἀγομένη 'Αθηνᾶς τῇ τρίτῃ. Photios, Suda, and the scholiast to *Il.* 8.39 claim the twenty-eighth day (τρίτη φθίνοντος) as Athena's birthday (see Day 28). Schmidt (*Geburtstag*, p. 99) and Jacoby (*FGrHist* IIIb Suppl., Vol. 1, p. 555) appear to accept both days as her birthday, but the fact that all the other "Olympian" deities had their birthdays in the first decade of the month would suggest that Athena's birthday must fall into this period. The variant tradition of the twenty-eighth day as her birthday may have arisen from the celebration of the Panathenaia on Hekatombaion 28. However that may be, the evidence *supra* is sufficient to establish that the third day of each month was devoted to Athena.

The scholiast to Aristophanes *Plut.* 1126 claims that this day was sacred to the Charites: ἔξω τῶν ἑορτῶν ἱεραί τινες τοῦ μηνὸς ἡμέραι νομίζονται 'Αθήνησι θεοῖς τισίν, οἷον νουμηνία καὶ ἑβδόμη 'Απόλλωνι, τετρὰς 'Ερμῇ καὶ ὀγδόη Θησεῖ· Χάρισι τρίτη. There was, of course, a cult of the Charites in Athens (Aristophanes *Thesm.* 300 and Pausanias 9.35.1-4), but this schoſion alone associates the third day of the month with them.

Day 4

The fourth day of each month was devoted to three deities: Herakles, Hermes, and Aphrodite. Eros also, through his association with Aphrodite, may have laid claim to this day. Two scholia briefly introduce the evidence for associating these deities with this day: schol. to Hesiod *Op.* 770, ἡ μὲν τετρὰς 'Ηρακλέους καὶ 'Ερμοῦ ἐστιν; schol. to Hesiod *Op.* 800, ἡ τετάρτη ἱερὰ 'Αφροδίτης καὶ 'Ερμοῦ.

That this day was Herakles' birthday is established by a proverb

MONTHLY FESTIVAL DAYS

entitled τετράδι γέγονας, a proverb which is quoted throughout the scholiastic and lexicographical tradition. It is stated most authoritatively in a citation of Philokhoros by the scholiast to Plato *Ap.* 19C:

Ἀριστώνυμος δ᾽ ἐν Ἡλίῳ ῥιγοῦντι καὶ Σανυρίων ἐν Γέλωτι τετράδι φασὶν αὐτὸν γενέσθαι· διὸ τὸν βίον κατέτριψε ἑτέροις πονῶν. οἱ γὰρ τετράδι γεννώμενοι πονοῦντες ἄλλοις καρποῦσθαι παρέχουσιν, ὡς καὶ Φιλόχορος ἐν τῇ πρώτῃ Περὶ ἡμερῶν ἱστορεῖ. ταύτῃ δὲ καὶ Ἡρακλῆ φησι γεννηθῆναι.

Jacoby (*FGrHist* 328 F85 and commentary) discusses the proverb and the ancient evidence concerning it (Photios and Suda τετράδι γέγονας, Suda τετραδισταί, Zenob. *Prov.* 6.7, Eustathius to *Od.* 5.262 and *Il.* 24.336). The sacrifice to the Herakleidai on Mounichion 4 on the Erkhia calendar (see Mounichion 4), the only sacrifice in the cult circle of Herakles on the calendar, confirms that this was Herakles' day.

The Homeric Hymn to Hermes (IV), line 19, establishes that this day was also Hermes' birthday, τετράδι τῇ προτέρῃ, τῇ μιν τέκε πότνια Μαῖα. This is confirmed by the scholia to Hesiod *Op.* 770 and 800 (cited *supra*), Plutarch *Mor.* 738F, and Eustathius to *Il.* 24.336. The Athenian honors paid to him on this day are suggested in the schol. to Aristoph. *Plut.* 1126 (ἡ τετρὰς ἐνομίζετο τοῦ Ἑρμοῦ· καὶ καθ᾽ ἕκαστον μῆνα ταύτῃ τῇ ἡμέρᾳ ἀπετίθεντο [πλακοῦντα] τῷ Ἑρμῇ), and are confirmed by a sacrifice to Hermes on the Erkhia calendar (see Thargelion 4). The perquisites for the herald were specified with unusual detail in the entry for this day.

Aphrodite was also born on this day, and thus the day was especially suitable for συνουσία: schol. to Hesiod *Op.* 800, ἡ τετάρτη ἱερὰ Ἀφροδίτης καὶ Ἑρμοῦ, καὶ διὰ τοῦτο πρὸς συνουσίαν ἐπιτηδεία. Athenaios 14.659D ff. cites a fragment of Menander (Kock, frag. 292) in describing a private celebration of the Tetradistai (a group of men who met on the fourth day of each month) in Aphrodite's honor:

διόπερ Μένανδρος ἐν Κόλακι τὸν τοῖς τετραδισταῖς
διακονούμενον μάγειρον ἐν τῇ τῆς πανδήμου Ἀφροδίτης
ἑορτῇ ποιεῖ ταυτὶ λέγοντα·
σπονδή. δίδου σὺ σπλάγχν᾽ ἀκολουθῶν. ποῖ βλέπεις;
σπονδή. φέρ᾽, ὦ παῖ Σωσία. σπονδή. καλῶς.
ἔγχει. θεοῖς Ὀλυμπίοις εὐχώμεθα

II. PROLEGOMENA TO CALENDAR

Ὀλυμπίαισι πᾶσι πάσαις (λάμβανε
τὴν γλῶτταν ἐπὶ τούτῳ) διδόναι σωτηρίαν,
ὑγίειαν, ἀγαθὰ πολλά, τῶν ὄντων τε νῦν
ἀγαθῶν ὄνησιν πᾶσι. ταῦτ' εὐχώμεθα.

Eros has been associated with this day, both because of his relationship with Aphrodite and because of Plato's statement in *Symposium* 203C, διὸ δὴ καὶ τῆς Ἀφροδίτης ἀκόλουθος καὶ θεράπων γέγονεν ὁ Ἔρως, γεννηθεὶς ἐν τοῖς ἐκείνης γενεθλίοις. The passage from the *Symposium* is, however, from Socrates' recollection of Diotima's story, and may represent the product of Plato's vivid imagination rather than mythological tradition. On the other hand, the private celebration in Eros' honor attested for the fifth century B.C. on this day in Mounichion (see Mounichion 4) would suggest, as Jacoby noted (*FGrHist* IIIb Suppl., Vol. 2, pp. 272–273, note 5), that Eros did have his claim to this day.

Thus three, or perhaps four, deities were honored on this day each month. According to Theophrastos, *Char.* 16, on this day the superstitious offered to the Hermaphrodites, the unique combination of Hermes and Aphrodite. As has been shown, numerous religious activities are attested for the deities who claimed this day, and the day must be viewed as a monthly festival day.

Day 6

Artemis' birthday was on Thargelion 6 (Diog. Laert. 2.44, Θαργηλιῶνος ἕκτῃ, ὅτε καθαίρουσιν Ἀθηναῖοι τὴν πόλιν καὶ τὴν Ἄρτεμιν Δήλιοι γενέσθαι φασίν). The monthly celebrations of the birthdays of other deities suggest the possibility of monthly celebrations of Artemis' birthday, and this is confirmed by Proklos *in Ti.* 200D, καὶ τὴν ἑβδόμην ἱερὰν ἔλεγον τοῦ θεοῦ... καθάπερ τὴν ἕκτην Ἀρτέμιδος. The celebration of the festival of Artemis Agrotera on Boedromion 6, and the procession to the Delphinion in Artemis' honor together with the meeting of the Soteriastai on Mounichion 6 establish that Artemis had some special claim to the sixth day in Athens. Artemis was commonly associated with the sixth day throughout the Greek world (Schmidt, *Geburtstag*, pp. 94–97). These various celebrations, both Athenian and non-Athenian, clearly indicate that the sixth day of each month in Athens was a festival day in Artemis' honor.

MONTHLY FESTIVAL DAYS

Day 7

The seventh day of each month was sacred to Apollo throughout the Greek world. Hesiod *Op.* 770-771 establishes that this was his birthday.

πρῶτον ἔνη τετράς τε καὶ ἑβδόμη ἱερὸν ἦμαρ·
τῇ γὰρ 'Απόλλωνα χρυσάορα γείνατο Λητώ.

For the numerous non-Athenian festivals of Apollo which fell on the seventh day, see Schmidt, *Geburtstag*, pp. 88-94, and Nilsson, *Die Entstehung und religiöse Bedeutung des griechischen Kalenders*[2], pp. 38-39. The scholiast to Aristoph. *Plut.* 1126 establishes that at Athens also the seventh day was sacred to Apollo: ἔξω τῶν ἑορτῶν ἱεραί τινες τοῦ μηνὸς ἡμέραι νομίζονται 'Αθήνησι θεοῖς τισίν, οἷον νουμηνία καὶ ἑβδόμη 'Απόλλωνι. This scholion is strikingly confirmed by a survey of the attested religious activities in Athens on the seventh day.

Hekatombaion 7	: Hekatombaia (?)
Metageitnion 7	: Salaminioi sacrifice to Apollo Patroos, Leto, Artemis, and Athena
Boedromion 7	: Boedromia (?)
Pyanopsion 7	: Pyanopsia
Gamelion 7	: Erkhians sacrifice to Apollo Delphinios and Apollo Lykeios
Thargelion 7	: Thargelia

The monthly religious activities devoted to Apollo on this day are specified by Proklos to Hesiod *Op.* 770 ff.: τὴν δὲ ἑβδόμην καὶ ὡς 'Απόλλωνος γενέθλιον ὑμνῶν, διὸ καὶ 'Αθηναῖοι ταύτην ὡς 'Απολλωνιακὴν τιμῶσι δαφνηφοροῦντες καὶ τὸ κανοῦν ἐπιστέφοντες καὶ ὑμνοῦντες τὸν θεόν. It is also noteworthy that only Apollo's festivals occurred on or included the seventh day of the month. The festival of no other deity was allowed to impinge upon this day.

Day 8

Proklos to Hesiod *Op.* 790 indicates that the eighth day of the month was sacred to Poseidon: τὴν ὀγδόην τοῦ μηνὸς τοῦ

II. PROLEGOMENA TO CALENDAR

Ποσειδῶνος ἱεράν. Plutarch *Thes*. 36 establishes that the Athenians paid honors to Poseidon on this day, καὶ γὰρ Ποσειδῶνα ταῖς ὀγδόαις [οἱ Ἀθηναῖοι] τιμῶσιν, and this is confirmed in part by an entry on an Athenian private sacrificial calendar of the first or second century A.D.:

IG II² 1367, lines 16–18:

Ποσιδεῶνος ᾗ ἱσταμένου πόπανον
χοινικιαῖον δωδεκόνφαλον καθήμεγ[ον]
Ποσιδῶνι χαμαιζήλῳ νηφάλιον

Several sources establish that Poseidon's son Theseus also received honors on this day each month in Athens:

Plutarch *Thes*. 36:

θυσίαν δὲ [οἱ Ἀθηναῖοι] ποιοῦσιν αὐτῷ [Θησεῖ] τὴν μεγίστην ὀγδόῃ Πυανεψιῶνος, ἐν ᾗ μετὰ τῶν ἠιθέων ἐκ Κρήτης ἐπανῆλθεν. οὐ μὴν ἀλλὰ καὶ ταῖς ἄλλαις ὀγδόαις τιμῶσιν αὐτόν

schol. to Aristoph. *Plut*. 627:

ταῖς ὀγδόαις τὰ Θησεῖα ἦγον καὶ ἀνεῖτο ἡ ὀγδόη πᾶσα τῷ Θησεῖ

schol. to Aristoph. *Plut*. 1126:

ἔξω τῶν ἑορτῶν ἱεραί τινες τοῦ μηνὸς ἡμέραι νομίζονται Ἀθήνησι θεοῖς τισίν, οἷον νουμηνία καὶ ἑβδόμη Ἀπόλλωνι, τετράς Ἑρμῇ καὶ ὀγδόη Θησεῖ

Hesychios ὀγδοαῖον (Latte, *Hesychios*, Vol. II, p. 733):

θυσία παρὰ Ἀθηναίοις τελουμένη Θησεῖ

This evidence establishes conclusively that this day at Athens was sacred to Poseidon and his son Theseus. But Poseidon and Theseus evidently did not maintain exclusive possession of this day, because on Elaphebolion 8 Asklepios received a sacrifice, and on Gamelion 8 the Erkhians sacrificed to Apollo Apotropaios, Apollo Nymphegetes, and the Nymphs.

MONTHLY FESTIVAL DAYS

Day 16

The discovery of a fragment of Philokhoros' Περὶ ἡμερῶν (Jacoby, *FGrHist* 328 F 86) has suggested that cakes were taken to the sanctuaries of Artemis and to the crossroads on this day each month.

> Φιλόχορος ἐν τῇ Περὶ ἡμερῶν ἓξ ἐπὶ δέκα· καὶ τοὺς καλουμένους δὲ νῦν ἀμφιφῶντας ταύτῃ τῇ ἡμέρᾳ πρῶτον ἐνόμισαν οἱ ἀρχαῖοι φέρειν εἰς τὰ ἱερὰ τῇ Ἀρτέμιδι καὶ ἐπὶ τὰς τριόδους.

See also Athenaios 14.645A. The Suda (ἀνάστατοι) and modern scholars have always associated these offerings only with the Mounichia of Mounichion 16, which was a state festival day. Philokhoros in the above fragment may have been discussing only the Mounichia of Mounichion 16 in his general discussion of the sixteenth day. But Jacoby (*FGrHist* IIIb Suppl., Vol. 1, pp. 369–370) persuasively argues that in addition to the Mounichia there were also monthly offerings to Artemis on the sixteenth day of every month. But, as Jacoby notes, there is no indication that the monthly offering was an ἑορτὴ δημοτελής.

Days 18 and 19

Proklos on Hesiod *Op.* 810 (Jacoby, *FGrHist* 328 F 190) creates a puzzle concerning these days:

> τὴν ἐννεακαιδεκάτην ὡς καὶ τὴν ὀκτωκαιδεκάτην τὰ πάτρια τῶν Ἀθηναίων καθαρμοῖς ἀποδίδωσι καὶ ἀποτροπαῖς, ὡς Φιλόχορος λέγει

Jacoby (*FGrHist* IIIb Suppl., Vol. 1, p. 555) assigns this fragment to Philokhoros' Περὶ ἡμερῶν, thereby indicating that the fragment refers to the eighteenth and nineteenth days of every month. There is nothing, however, in the Athenian calendar of festivals to indicate that these days were specially devoted to purifications and apotropaic rites. It may be possible that this fragment should be assigned to Philokhoros' Περὶ ἑορτῶν, and in fact refers to Boedromion 18 and 19. Boedromion 19, the day of the procession to Eleusis for the Mysteries, provided numerous apotropaic rites, and some such rites may have been performed on Boedromion 18, a day for which our evidence is ambiguous (see Boedromion—Summary).

II. PROLEGOMENA TO CALENDAR

Days 27, 28, and 29

Some scholars[13] have claimed that days 27, 28, and 29 of every month were ἀποφράδες on the basis of *Etym. Magn.* 131.13 ff.:

ἀποφράδας ἔλεγον οἱ Ἀττικοὶ τὰς ἀπηγορευμένας ἡμέρας, ἃς ὑπελάμβανον χείρους εἶναι τῶν ἄλλων· ἃς δὴ καὶ ἐπεικάδας καλοῦσι φθίνοντος τοῦ μηνὸς τετράδα, τρίτην, δευτέραν. ἢ τὰς ἡμέρας ἐν αἷς τὰς φονικὰς δίκας ἐδίκαζον.

ἡμέραι ἀποφράδες were not simply "festival days," and must not be equated with the Roman *dies nefasti*.[14] ἡμέραι ἀποφράδες were rather a special class of festival days. These days were thought to be "polluted" and "impure" (οὐ καθαραί). The day of the annual Plynteria is an attested example of an Athenian ἡμέρα ἀποφράς. The general mood of ἡμέραι ἀποφράδες differed significantly from that of most Attic festival days.

The *Etym. Magn.* would lead us to believe that days 27, 28, and 29 of every month were ἀποφράδες. But the procession of the Panathenaia on Hekatombaion 28, and the celebration of the Theogamia on Gamelion 27 were certainly *not* ἡμέραι ἀποφράδες, and thus they indicate that the general statement of *Etym. Magn.* is not correct.

The statement in *Etym. Magn.* is thus in error, and the source of this error can be, I believe, established. As so often happens in lexicographical sources, the statement has been distorted by careless and inaccurate compilation. The opening statement (ἀποφράδας ἔλεγον...τῶν ἄλλων) is essentially correct. The following two statements (ἃς δὴ καὶ...δευτέραν, and ἢ...ἐδίκαζον), I propose, stem from two sources which discussed the same subject, viz. the days on which the Areopagos council judged homicide cases. One source stated that the days on which the Areopagos council judged homicide cases were ἀποφράδες. A second source enumerated the specific days on which the Areopagos council could judge such cases, i.e., the twenty-seventh, twenty-eighth, and twenty-ninth days of a month. That these were the meeting days of the Areopagos council for this purpose is established by Pollux 8.117: Ἄρειος

---

[13] E.g., M. P. Nilsson, *Die Entstehung und religiöse Bedeutung des griechischen Kalenders*², p. 42.

[14] See the forthcoming discussion of ἡμέραι ἀποφράδες by author in *AJP*.

MONTHLY FESTIVAL DAYS

πάγος...καθ' ἕκαστον δὲ μῆνα τριῶν ἡμερῶν ἐδίκαζον ἐφεξῆς, τετάρτῃ φθίνοντος, τρίτῃ, δευτέρᾳ. The compiler of the *Etym. Magn.*, however, reversed the proper sequence of these two statements, and thereby gave the impression that Days 27, 28, and 29 were always ἀποφράδες. If the above interpretation of the *Etym. Magn.* citation is correct, the following two conclusions should be drawn: a day on which the Areopagos council judged a homicide case was ἀποφράς, a belief that is consistent with Greek religious practices; and, secondly, the Areopagos council could judge such cases only on the twenty-seventh, twenty-eighth, and twenty-ninth days of a month.

The Areopagos council would meet, of course, only when cases arose. If no cases arose in one month, then no one of these days was ἀποφράς. A day would be ἀποφράς only if the court were in session. If the court met only on one or two days, only these one or two days would be ἀποφράς. The court obviously would not be held on days of major festivals such as the Panathenaia.

The conclusion then is that these days were not ἀποφράδες *per se*. If, however, the Areopagos court were trying a case on one of these days, then that day became ἀποφράς.

Photios (and Suda) claims Day 28 as Athena's birthday: τριτογενής· ἡ 'Αθηνᾶ· ἤτοι ὅτι...τρίτῃ φθίνοντος [ἐγεννήθη] ὡς καὶ 'Αθηναῖοι ἄγουσιν. This is also stated by the scholiast to *Il.* 8.39, τριτογένεια· ἢ ὅτι τρίτῃ φθίνοντος ['Αθηνᾶ] ἐτέχθη. Harpokration and *Etym. Magn.* 767.40 ff., however, give Day 3 as her birthday (see Day 3). There is no indication that the Athenians celebrated a monthly festival on every twenty-eighth day in honor of Athena. Her birthday, as those of the other "Olympian" deities, properly belongs in the first decade of the month (see Day 3), and thus should be associated with the third day of the month. The three sources which give Day 28 evidently stem from a common source which attempted to link the epithet τριτογενής with the day of the celebration of the Panathenaia, Hekatombaion 28.

Day 30

Apollodoros, as quoted in Athenaios 7.325A (*FGrHist* 244 F 109), states that δεῖπνα were given to Hekate on the thirtieth day of each month: τῇ δὲ Ἑκάτῃ ἀποδίδοται ἡ τρίγλη διὰ τὴν τῆς ὀνομασίας

II. PROLEGOMENA TO CALENDAR

κοινόνητα· τριοδῖτις γὰρ καὶ τρίγληνος, καὶ ταῖς τριακάσι δ' αὐτῇ τὰ δεῖπνα φέρουσι. This practice is specified in more detail by the scholiast to Arist. *Plut.* 594, κατὰ δὲ νουμηνίαν οἱ πλούσιοι ἔπεμπον δεῖπνον ἑσπέρας, ὥσπερ θυσίαν τῇ Ἑκάτῃ ἐν ταῖς τριόδοις· οἱ δὲ πένητες ἤρχοντο πεινῶντες, καὶ ἤσθιον αὐτὰ καὶ ἔλεγον ὅτι ἡ Ἑκάτη ἔφαγεν αὐτά. The scholiast was evidently following the practice of strict time reckoning, whereby the day began at sunset, and thus the offering which was made in general terms on the evening of Day 30 could be designated κατὰ νουμηνίαν. There is no indication that these private offerings in the evening entailed a monthly festival day.

Summary of Monthly Festival Days

The survey *supra* has established for Athens the following monthly festival days:

Day 1	Noumenia
Day 2	Day of the ἀγαθὸς δαίμων
Day 3	Athena's birthday
Day 4	Day devoted to Herakles, Hermes, Aphrodite, and Eros
Day 6	Artemis' birthday
Day 7	Apollo's birthday
Day 8	Day sacred to Poseidon and Theseus

These are the only monthly festivals which are positively attested for Athens. These festival days obviously dominate the first decade of the month, the decade of the "waxing moon."

CHAPTER III
CALENDAR OF THE ATHENIAN YEAR

HEKATOMBAION

Hekatombaion 1

This day was a monthly festival day—the Noumenia.

Hekatombaion 2

This day was the monthly festival day of the Agathos Daimon. For a possible financial transaction on this day, refer to Hekatombaion 2 in Appendix I.

Hekatombaion 3

This day was a monthly festival day devoted to Athena. See also Hekatombaion 3 in Appendix II.

Hekatombaion 4

This was a monthly festival day devoted to Herakles, Hermes, Aphrodite, and Eros. On this day in 329/8 B.C. a work crew began their assignment at Eleusis.

IG II² 1672, lines 32-34:

 μισθωτοῖς τοῖς ἐργασαμέν-
οις ἐν τῶι ἱερῶι, ἀνδράσιν δέκα, ἀπὸ τῆς τετράδος ἱσταμένου τοῦ Ἑκατονβαιῶνο : τῶι ἀνδρὶ οἰκοσίτωι :⊢ΙΙΙ
ἡμερῶν : ΔΔΔΔ : κεφά : ⌐ Η : ἄχρι τῆς τρίτης ἐπὶ δέκα τοῦ
Μεταγειτο :

The men then worked every day until Metageitnion 13, forty days in all.

III. THE ATHENIAN CALENDAR

Hekatombaion 5

No evidence as to the nature of this day survives.

Hekatombaion 6

This was a monthly festival day devoted to Artemis.

Hekatombaion 7

This day was probably the day of the Hekatombaia, the annual festival of Apollo Hekatombaios. The month is established by *Etym. Magn.* 321.5, Ἑκατομβαιὼν δὲ ὠνόμασται διὰ τὰς τοῦ Ἀπόλλωνος θυσίας· θύουσι γὰρ αὐτῷ Ἑκατομβαίῳ, and Bekker's *Anecd.* 1.247.1, Ἑκατομβαιών· ὠνομάσθη δὲ οὕτως, ἐπειδὴ ἱερός ἐστι τοῦ Ἀπόλλωνος. The day is suggested by the fact that the seventh day of the month was sacred to Apollo, and by the fact that the sacrifice to Apollo Hekatombaios on Mykonos occurred on Hekatombaion 7.[1]

Restoration to give Hekatombaion 7:

A sacrifice to Apollo Apotropaios on the sacred calendar of the Marathonian Tetrapolis, *IG* II² 1358, Column I, lines 24–26.

Hekatombaion 8

This was a monthly festival day devoted to Poseidon and Theseus. The private religious association of Thracian orgeones concerned with the Bendideia held a meeting on this day in 263/2 B.C.

IG II² 1283, lines 2–3:

Ἐπὶ Πολυστράτου ἄρχοντος μηνὸς Ἑκατομβαιῶνος ὀγδόη-
ι ἱσταμένου· ἀγορᾶι κυρίαι.

Hekatombaion 9

No evidence as to the nature of this day survives.

[1] Dittenberger, *Sylloge*³ 1024, lines 29–30: Ἑκατομβαιῶνος ἑβδόμηι ἱσταμένου Ἀπόλλωνι Ἑκατομβίωι ταῦρος καὶ δέκα ἄρνες.

HEKATOMBAION

Hekatombaion 10

No indisputable evidence as to the nature of this day survives.

Restoration to give Hekatombaion 10:
A meeting of the deme of the Ikarians, *Hesp* 1948, pp. 142-143 (D. M. Robinson), lines 4-5.

Hekatombaion 11

Four meetings of the Ekklesia establish this day as a meeting day of the Ekklesia.

Demosthenes 24.26: 352 B.C.

ἀλλὰ τῆς ἐκκλησίας, ἐν ᾗ τοὺς νόμους ἐπεχειροτονήσατε, οὔσης ἑνδεκάτῃ τοῦ Ἑκατομβαιῶνος μηνός

IG II² 365, lines 1-4: 323/2 B.C.

Ἐπὶ Κηφισοδώρου ἄρχον[τος ἐπὶ τῆς Ἱππο]-
θωντίδος πρώτης πρυτα[νείας ἧι Εὐκλῆς]
[Π]υθοδώρου ἐγραμμάτευε[ν· Ἑκατομβαιῶν]-
[ο]ς ἑνδεκάτει, (ἑνδεκάτει) τῆς πρυτανεί[ας]

IG II² 650, lines 1-5: 286/5 B.C.

Ἐπὶ Διοκλέους [ἄρχοντος ἐπὶ] τῆς [.......]
ίδος πρώτης π[ρυτανείας ἧι] Ξενοφῶ[ν Νικέ]-
ου Ἁλαιεὺς ἐγ[ραμμάτε]υ[εν· Ἑκ]ατομβ[αιῶνο]-
ς ἑνδεκάτει, ἑ[νδεκάτει] τῆ[ς πρ]υτανείας· ἐκκλησία

Hesp 1969, pp. 418-425, no. 1 254/3 B.C.
(J. S. Traill), lines 1-4:

Ἐπὶ Φιλίνου ἄρχοντος ἐπὶ τῆς Αἰγεῖδος πρώτης
   πρυτανείας ἧ-
ι Θεότιμος Στρατοκλέους Θοραιεὺς ἐγραμμάτευεν·
   Ἑκατομ-
βαιῶνος ἑνδεκάτει, ἑνδεκάτει τῆς πρυτανείας·
   ἐκκλησία κυ-
ρία

III. THE ATHENIAN CALENDAR

Restoration to give Hekatombaion 11:

A meeting of the Ekklesia in 307/6 B.C., Pritchett and Meritt, *Chronology*, pp. 7-8, lines 1-6.

Hekatombaion 12

Demosthenes, 24.26, establishes this day as the day of the Kronia, and thus a festival day:

[Τιμοκράτης] οὔτε γὰρ ἐξέθηκε τὸν νόμον, οὔτ' ἔδωκεν εἴ τις ἐβούλετ' ἀναγνοὺς ἀντειπεῖν, οὔτ' ἀνέμεινεν οὐδένα τῶν τεταγμένων χρόνων ἐν τοῖς νόμοις, ἀλλὰ τῆς ἐκκλησίας, ἐν ᾗ τοὺς νόμους ἐπεχειροτονήσατε, οὔσης ἑνδεκάτῃ τοῦ Ἑκατομβαιῶνος μηνός, δωδεκάτῃ τὸν νόμον εἰσήνεγκεν, εὐθὺς τῇ ὑστεραίᾳ, καὶ ταῦτ' ὄντων Κρονίων καὶ διὰ ταῦτ' ἀφειμένης τῆς βουλῆς, διαπραξάμενος μετὰ τῶν ὑμῖν ἐπιβουλευόντων καθέζεσθαι νομοθέτας διὰ ψηφίσματος ἐπὶ τῇ τῶν Παναθηναίων προφάσει.

He here also states that in 352 B.C. a meeting of the Ekklesia occurred on this day, but he intimates that this was highly irregular. The Boule had been dismissed because of the festival, and, as Demosthenes here suggests, Timokrates in a stealthy and underhanded manner arranged that the Ekklesia, attended by his supporters, met on this day and passed into law his proposal.[2] Demosthenes does not specify his complaint about a meeting on this day; he does not charge Timokrates with παρανομία or ἀσέβεια. But his suggestion that it was an irregular procedure is unmistakable.

Hekatombaion 13

No evidence as to the nature of this day survives.

Hekatombaion 14

Except for a financial transaction (see Hekatombaion 14 in Appendix I), no evidence as to the nature of this day survives.

[2] For a similar charge see Aeschines 3.66–67 (cited under Elaphebolion 8).

HEKATOMBAION

Hekatombaion 15

A biennial sacrifice is recorded for this day on the major fragment of the Athenian State Sacrificial Calendar (J. H. Oliver, *Hesp* 1935, no. 2, p. 21), lines 30–43.

τάδε τὸ ἕτερον ἔτος θύεται ᾿Α [- - - - -]
   Ἑκατομβαιῶνος
     πέμπτηι ἐπὶ δέκα
     ἐκ τῶν φυλο-
     βασιλικῶν
       Γλεόντων φυλῆι
       Λευκοταινίων
       τριττύι οἶν
ΙΙΙ   λειπογνώμονα
ΙΙΙΙΙ   ἱερεώ[σ]υνα
       φυλοβ[α]σιλεῦσι
Ι   νῶτο
       κήρυκι ὦμο
ΙΙΙΙ   ποδῶν κεφαλῆς

This sacrifice is to be associated with the festival of the Synoikia on the following day, Hekatombaion 16. The sacrifices on Hekatombaion 16 are authorized by the same source (ἐκ τῶν φυλοβασιλικῶν), and the same tribesmen (Gleontes) are prominent. Hekatombaion 16 is the more important day, for the sacrifices on that day are larger. No source suggests that the Synoikia were a two-day festival, but the above sacrifice establishes that there was at the least a prefestival day for the Synoikia.

No public meetings are attested for Hekatombaion 15. Legal proceedings, however, are attested to have occurred on this day ca. 330 B.C.

IG II² 1578, lines 1–2:

[πολεμαρχοῦν]τος Δημοτέλους τοῦ ᾿Αντ[ι]μάχου ᾿Αλ[α]-
[ιέως δίκαι ἀπο]στασίου Ἑκατονβαιῶνος π[έμπτ]ει ἐπὶ
[δ]έ[κα]

Hekatombaion 16

This day was the day of the celebration of the Synoikia in Athens. Thucydides, 2.15, confirms the celebration of the festival, καὶ

III. THE ATHENIAN CALENDAR

Ξυνοίκια ἐξ ἐκείνου (the time of the Synoikism) Ἀθηναῖοι ἔτι καὶ νῦν τῇ θεῷ ἑορτὴν δημοτελῆ ποιοῦσιν, and Plutarch, *Thes.* 24.4, giving the name in an altered form, provides the date, (Theseus) ἔθυσε δὲ καὶ Μετοίκια τῇ ἕκτῃ ἐπὶ δέκα τοῦ Ἑκατομβαιῶνος, ἣν ἔτι νῦν θύουσι. The Athenian State Sacrificial Calendar records a biennial state sacrifice on this day.

lines 44–58:

        (τάδε τὸ ἕτερον ἔτος θύεται Ἀ [- - - - -]) (30)
        (Ἑκατομβαιῶνος) (31)
        ἕκτηι ἐπὶ δέκα
        ἐκ τῶν φυλο-
        βασιλικῶν
        Γλεόντων φυλῆι
        Διὶ Φρατρίωι καὶ
        Ἀθηναίαι Φρα-
        τρίαι βόε δύο
𐅀   [λ]ειπογνώμονε
: Δ Γ⊢ ἱερεώσυνα
        φυλοβασιλεῖ
        σκέλος
        κήρυ[κ]ι χέλυος
⊢⊢ΙΙΙ   ποδ[ῶν] κεφαλῆς
        τ [. . . .] ει κριθῶν
        μ[έδιμ]νοι

Oliver (p. 26) correctly associated this sacrifice with the Synoikia. Thucydides and Plutarch give the impression that the festival was annual, and it is surprising to find it listed as biennial in the State Calendar. Recent unpublished studies by Dow have indicated that the biennial sacrifices for the Synoikia as listed on the State Calendar formed only a part of the state festival. The Synoikia were certainly celebrated annually, because the other festivals associated with the Theseus legend (Delphinia, Oskhophoria, Panathenaia, and Pyanopsia) were all annual. These sacrifices from which the Gleontes alone received a banquet were perhaps made biennial by Nikomakhos' reforms. The absence of any attested public meetings on this day may confirm that the Synoikia were celebrated annually.

The scholiast to Aristophanes *Pax* 1019, φασὶ γὰρ τῇ τῶν Συνοικίων (emended from συνοικείων and συνοικεσίων) ἑορτῇ θυσίαν τελεῖσθαι Εἰρήνῃ, τὸν δὲ βωμὸν μὴ αἱματοῦσθαι, Ἑκατομβαιῶνος μηνὸς ἕκτῃ ἐπὶ δέκα, associates a sacrifice to Eirene with the Synoikia on this day. The value of the scholion is problematical,[3] but it evidently refers to the sacrifice established in 374 B.C.[4] It appears that this sacrifice, perhaps in some way combined with the Synoikia, was not biennial. According to the accounts of the treasurers of Athena, *IG* II² 1496, there was sizable income from this sacrifice in two successive years, 333/2 B.C. (lines 93-95), and 332/1 B.C. (lines 126-128).

Restoration to give Hekatombaion 16:

A meeting of the Ekklesia in 332/1 B.C., *IG* II² 420, lines 1-4 as suggested by Meritt, *AJP* 85 (1964), pp. 304-306.

Hekatombaion 17 and 18

Sacrifices are specified for these days in the early third century B.C. by the religious association of the Orgeones.

Hesp 1942, pp. 282-287, no. 55
(Meritt), lines 12-16:

ἔδοξεν τοῖς ὀργεῶσιν· τὸν ἑστιάτορα θύειν τὴν [θυσί]-
αν μηνὸς Ἑκατονβαιῶνος ἑβδόμει καὶ ὀγδόει ἐπ[ὶ δ]-
έκα· θύειν δὲ τῆι πρώτει ταῖς ἡρωίναις χοῖρον, τῶι δὲ[ἥ]-
[ρ]ωι ἱερεῖον τέλεον καὶ τράπεζαν παρατιθέναι, τὲι δ[ὲ]
[ὑστερ]άαι τῶι ἥρωι ἱερεῖον τέλεον

These sacrifices are performed by a private association, and do not suggest a public festival. No other information concerning the nature of these days survives.

Hekatombaion 19

No evidence as to the nature of this day survives.

---

[3] Deubner, *Feste*, pp. 37-38.
[4] Isokrates 15.109-110, Deubner, *Feste*, p. 38, and Jacoby, *FGrHist* IIIb Suppl., Vol. 1, pp. 523-526.

III. THE ATHENIAN CALENDAR

Hekatombaion 20

Except for financial transactions (see Hekatombaion 20 in Appendix I), no evidence as to the nature of this day survives.

Hekatombaion 21

Sacrifices to Kourotrophos and Artemis are prescribed for this day on the sacrificial calendar of the deme Erkhia.[5]

Column Γ, lines 1–12
['Ε]κατομβαιῶν-
ος δεκάτει ὑ-
στέραι, Κουρ-
οτρόφωι, ἐς Σ-
ωτιδῶν Ἐρχι,
χοῖρος, οὐ φο-
ρά,⊢⊢⊢
Ἀρτέμιδι ἐς
Σωτιδῶν Ἐρχ-
ι: αἴξ, οὐ φορά,
τὸ δέρμα κατ-
αγίζ: Δ

Column Δ, lines 1–12
Ἑκατομβαιῶν-
ος δεκάτει ὑ-
στέραι, Κορο-
[τρ]όφωι, ἐπὶ τ-
[ὸ] Ἄκρο Ἐρχιᾶ,
χοῖρος, οὐ φο-
ρά ⊢⊢⊢
Ἀρτέμιδι, ἐπ-
ὶ τὸ Ἄκρο Ἐρχ-
ιᾶ, αἴξ, οὐ φορ-
ά, δέρμα κατα-
ιγίζε: Δ

These sacrifices performed by a deme do not by themselves presuppose a state festival day, and no other evidence concerning the nature of this day survives.

Hekatombaion 22

A meeting of the Boule in 411 B.C. establishes this day as a meeting day of the Boule.

IG I² 298, lines 14–17:

ψηφισαμέ-
νης τῆς βολῆς Ἑκατ-
[ο]μβαιῶνος ἐνάτηι
[φθί]νοντος

[5] G. Daux, *BCH* 87 (1963), pp. 603–634.

32

HEKATOMBAION

Hekatombaion 23

On this day in 261/0 B.C. there was a meeting of a private religious association.

IG II² 1282, lines 2–3:

Ἐπ' Ἀντιπ[ά]τρου ἄρχοντος, Ἑκα[το]νβαιῶ-
νος ὀγδόει μετ' εἰκάδας, ἀγορᾶ[ι κ]υρ[ί]αι

Refer also to Hekatombaion—Summary.

Restoration to give Hekatombaion 23:
A meeting of the Boule in 423/2 B.C., *IG* I² 324, lines 57–59 as restored by Meritt, *AFD*, p. 132. But only a few letters of these lines are readable, and no restoration can be certain.[6]

Hekatombaion 24

No evidence as to the nature of this day survives, but refer also to Hekatombaion—Summary.

Hekatombaion 25

No evidence as to the nature of this day survives, but refer also to Hekatombaion—Summary.

Hekatombaion 26

No evidence as to the nature of this day survives, but refer also to Hekatombaion—Summary.

Hekatombaion 27

No evidence as to the nature of this day survives, but refer also to Hekatombaion—Summary.

[6] Pritchett and Neugebauer, *Calendars*, p. 105, "Our position, therefore, is that although *IG* I² 324 may be restored according to a regular pattern in the length of prytanies the preserved portions are the only part which may be used for historical evidence."

III. THE ATHENIAN CALENDAR

Hekatombaion 28

Proklos, *in Ti.* 9B, τὰ γὰρ μεγάλα [Παναθήναια] τοῦ Ἑκατομβαιῶνος ἐγίνετο τρίτῃ ἀπιόντος, and the scholiast to Plato *Rep.* 327A, καὶ ταῦτα [τὰ μεγάλα Παναθήναια] μὲν ἦγον εἰς ἄστυ Ἑ[κα]τομβαιῶνος μηνὸς τρίτῃ ἀπιόντος, establish this day as the "Haupttag" of the Panathenaia. On this day occurred the night festival, the procession from the Kerameikos, the presentation of the peplos to Athena, and the major state sacrifices (Deubner, *Feste*, pp. 24–35).

Hekatombaion 29

No evidence as to the nature of this day survives, but refer also to Hekatombaion—Summary.

Hekatombaion 30

No indisputable evidence as to the nature of this day survives. Refer also to Hekatombaion—Summary and to Hekatombaion 30 in Appendix II.

Hekatombaion—Summary

The number of days included in the celebration of the Panathenaia has never been established. Some ancient sources[7] suggest three or four days, but Aristides, *Panathen.* 147, indicates that a day or more might be added if it seemed necessary. Mommsen (*Feste*, p. 153) assigns nine days, Hekatombaion 21–29, to the festival and its contests. The last two days of a month were regularly meeting days. Every month but Hekatombaion has at least one meeting attested on either the twenty-ninth or thirtieth day. Because there is no such meeting in Hekatombaion, it may be suggested that the Panathenaia included Hekatombaion 29 and 30. The lack of all public meetings in the last eight days may suggest that the Panathenaia usually included Hekatombaion 23–30.

[7] Schol. to Aristides *Panathen.* 147, and schol. to Euripides *Hec.* 469.

METAGEITNION

Metageitnion 1
This day was a monthly festival day—the Noumenia.

Metageitnion 2
This day was a monthly festival day of the Agathos Daimon. A meeting of the Boule in 203/2 B.C. also establishes this day as a meeting day of the Boule.

IG II² 915 as re-edited by Dow, *Prytaneis*, pp. 89-91, no. 40, lines 16-19:

Ἐπὶ Προξενίδου ἄρχοντος ἐπὶ τῆς Ἱπποθωντίδος
 δευτέρα[ς πρυ]-
τανείας ἧι Εὔβουλος Εὐβουλίδ[ο]υ Αἰξωνεὺς ἐγραμμάτευ[εν·]
Μεταγειτνιῶνος δευτέραι ἱσταμένου, πέμπτηι τῆς πρυ[τα]-
νείας· βουλὴ ἐμ βουλευτηρίωι

According to [Demosthenes] 42.5 τοῦ γὰρ Μεταγειτνιῶνος μηνός, ὦ ἄνδρες δικασταί, τῇ δευτέρᾳ ἱσταμένου ἐποίουν οἱ στρατηγοὶ τοῖς τριακοσίοις τὰς ἀντιδόσεις, legal proceedings occurred on this day in the fourth century B.C. This speech, though assumed not to have been written by Demosthenes, is a genuine speech from the fourth century B.C.,[8] and thus the calendric information may be considered valid.

Metageitnion 3
This day was a monthly festival day devoted to Athena.

Metageitnion 4
This day was a monthly festival day devoted to Herakles, Hermes, Aphrodite, and Eros. A meeting of the Boule in 155/4 B.C. also establishes this day as a meeting day for the Boule.

---
[8] Louis Gernet, *Démosthène, Plaidoyers Civils*, Vol. 2, Paris, 1957, pp. 76-77.

III. THE ATHENIAN CALENDAR

Hesp 1934, pp. 31-35, no. 21 (Meritt) as re-edited with new fragments by Dow, *Prytaneis*, pp. 148-153, no. 84, lines 42-45:

Ἐπὶ Μν[ησ]ιθέου ἄρχο(ντος) ἐπὶ τῆς Ἱπποθωντίδος
δευτέρας πρυ-
τανείας ἧι Φιλίσκος Κράτητος Παιανιεὺς ἐγραμμάτευεν·
Με-
ταγειτνιῶνος τετράδι ἱσταμένου, τετάρτει τῆς
πρυτ[ανε]ίας·
βουλὴ ἐμ βουλευτηρίωι

Metageitnion 5

No evidence as to the nature of this day survives.

Metageitnion 6

This was a monthly festival day devoted to Artemis.

Metageitnion 7

On this day the genos of the Salaminioi sacrificed to Apollo Patroos, Leto, Artemis, and Athena.

Hesp 1938, pp. 3-5 (W. S. Ferguson), lines 89-90:

Μεταγειτνιῶνος. ἑβδόμει Ἀπόλλωνι Πατρώιωι ὗν: ΔΔΔΔ,
Λητδι χοῖρο[ν]
[⊢]⊢⊢|||, Ἀρτέμιδι χοῖρον ⊢⊢⊢||| , Ἀθηνᾶι ἀγελάαι χοῖρον
⊢⊢⊢||| · ξύλα ἐφ᾽ ἱεροῖς καὶ εἰς τἄλλα ⊢⊢⊢||| .

This sacrifice further establishes that the seventh day of the month was devoted to Apollo. There may be in this sacrifice a reflection of the state Metageitnia (Plutarch *Mor.* 601B), which was celebrated in Metageitnion in honor of Apollo.[9]

[9] Harpokration (and Suda) Μεταγειτνιών· ἐν δὲ τούτῳ Ἀπόλλωνι Μεταγειτνίῳ θύουσιν.

METAGEITNION

Metageitnion 8

This was a monthly festival day devoted to Poseidon and Theseus. For a financial transaction on this day, see Metageitnion 8 in Appendix I.

Metageitnion 9

Three meetings establish this day as a meeting day of the Ekklesia.

IG II² 338, lines 1–5: 333/2 B.C.

Ἐπὶ Νικοκράτους ἄρχοντος ἐπὶ τῆς Αἰγηίδος
πρώτης πρυτανείας ἧι Ἀρχέλας Χαιρίου Παλ-
ληνεὺς ἐγραμμάτευεν· Μεταγειτνιῶνος ἐνάτηι ἱσταμένου, ἐνάτηι καὶ τριακοστῆι τῆς
πρυτανείας

Hesp 1954, pp. 287–296 (Dinsmoor), 271/0 B.C.
lines 2–6:

Ἐπὶ Πυθαράτου ἄρχοντος ἐπὶ τῆς Ἀντιγονίδος δευτέρας πρυτανείας ἧι Ἰσήγορος Ἰσοκράτου Κεφαλῆθεν ἐγραμμάτευεν· Μεταγειτνιῶνος ἐνάτει ἱσταμένου, ἑβδόμει τῆς πρυτανεί-
ας· ἐκκλησία κυρία

IG II² 687, lines 2–5: 265/4 B.C.

Ἐπὶ Πειθιδήμου ἄρχοντος ἐπὶ τῆς Ἐρεχθεῖδος δευτέρας π-
[ρ]υτανείας·
Μεταγειτνιῶνος ἐνάτει ἱσταμένου, ἐνάτει τῆς πρυτανεί-
ας· ἐκκλησία κυρία

Restorations to give Metageitnion 9:

A meeting of the Ekklesia in 307/6 B.C., *IG* II² 455, lines 1–4. These lines are restored by Pritchett and Meritt (*Chronology*, p. 20) to give Skirophorion 4, and by Meritt (*Year*, pp. 177–178) to give Thargelion 2.

A meeting of the Ekklesia at the end of the second century B.C., *IG* II² 1019, lines 1–3.

III. THE ATHENIAN CALENDAR

Metageitnion 10

No evidence as to the nature of this day survives.

Metageitnion 11

A meeting of the Ekklesia in 204/3 B.C. establishes this day as a meeting day of the Ekklesia.

IG II² 973, lines 2–6:

Ἐπὶ Ἀπολλοδώρου ἄρχοντος [ἐπὶ τῆς Πανδιο]-
νίδος δευτέρας πρυτανεία[ς ἦι....¹⁰....]
νος Ὀῆθεν ἐγραμμάτευεν· [Μεταγειτνιῶνος]
[ἑ]νδεκάτῃ, ἑνδεκάτει τ[ῆς πρυτανείας· ἐκκλη]-
[σί]α κυρία ἐν τῶι θε[άτρωι]

Restoration to give Metageitnion 11:
A meeting of the Ekklesia in 108/7 B.C., *IG* II² 1036, lines 7–9.

Metageitnion 12

A meeting in 250/49 B.C. establishes this day as a meeting day of the Ekklesia.

IG II² 778, lines 1–5:

Ἐπὶ Θερσιλόχου ἄρχοντος ἐ[πὶ τῆς...vτί]-
[δ]ος δευτέρας πρυτανείας ἦ[ι Διόδοτος Δ]-
ιογνήτου Φρεάρριος ἐγραμμ[άτευεν· Μετα]-
γειτνιῶνος δωδεκάτηι, δωδε[κάτηι τῆς πρ]-
υτανείας· ἐκκλησία κυρία

The sacrificial calendar of the deme Erkhia prescribes sacrifices on this day to Apollo Lykeios, Demeter, Zeus Polieus, and Athena Polias.

Column A, lines 1–5
Μεταγειτνιῶ-
νος δωδεκάτε-
ι, Ἀπόλλωνι Λ-
υκείωι, ἐν ἄστ-
ει, οἶς, οὐ φο, Δ ⊢⊢

Column B, lines 1–5
Μεταγειτνιῶ-
νος δω[δεκ]άτ-
ει, ἐν Ἐλευσι
ἐν ἄστει, Δήμ-
ητρι, οἶς, Δ

METAGEITNION

Column Γ, lines 13-18
[Μ]εταγειτνιῶ-
νος δωδεκάτ-
ει, Διὶ Πολιε͂,
ἐμ Πόλε ἐν ἄσ-
τε, οἷς, οὐ φορ-
ά, Δ ⊦⊦

Column Δ, lines 13-17
Μεταγειτνιῶ-
νος δωδεκάτ-
ε, Ἀθηνᾶι Πολ-
ιάδι, ἐμ Πόλε
ἐν ἄστε, οἷς: Δ

These sacrifices are to be performed at Eleusis and in Athens, and thus may suggest a state festival. There is, however, no other evidence to support this suggestion.

Metageitnion 13

No evidence as to the nature of this day survives, but refer also to Metageitnion—Summary.

Metageitnion 14

After ca. 163 B.C. the Technites of Dionysos celebrated the birthday of King Ariarathes V on this day.

IG II^2 1330, lines 30-34:

[θῦσαι ὑπὲρ σωτηρίας τῆς]
συνόδου καὶ βασιλέως Ἀριαράθου καὶ βασιλ[ίσσης]
[Νύσης - - - - - - - - - -]
καὶ μερίδα νεῖμαι πᾶ[σ]ιν τοῖς μετέχουσιν [τῆς συνόδου]
[καὶ παισὶ καὶ γυναιξὶν (?)]
[α]ὐ[τ]ῶν. μερίσαι δὲ τὸν [ἐπ]ιμελητὴν τοῦ Μετ[αγειτνιῶνος]
[μηνὸς τὴν τετράδα]
[ἐ]πὶ δέκα ὑπὲρ τοῦ βασ[ιλ]έως καὶ τὴν πέμπτ[ην ἐπὶ δέκα]
[ὑπὲρ τῆς βασιλίσσης]

There is no indication that this was a state festival. Refer also to Metageitnion—Summary.

Restoration to give Metageitnion 14:

A meeting of the Ekklesia in 125/4 B.C., *Hesp* 1933, pp. 163-165, no. 9 (Meritt), lines 1-3. Meritt later (*Year*, pp. 190-191) restored these lines to give Metageitnion 17.

39

III. THE ATHENIAN CALENDAR

Metageitnion 15

After ca. 163 B.C. the Technites of Dionysos celebrated on this day the birthday of Nysa, wife of King Ariarathes V. This is established by *IG* II² 1330, lines 30–34 (cited *supra* for Metageitnion 14). A private sacrifice is prescribed for this day in the first or second century A.D.

IG II² 1367, lines 1–3:

Μεταγιτνιῶνος θεαῖς β [...⁸⁻¹⁰...]
του τῆς παντελείας πόπανον [δωδεκόμ]-
φαλον χοινικιαῖον ΙΕ νηφάλιον

This sacrifice may be a reflection of the state Eleusinia. Refer to Metageitnion—Summary.

Metageitnion 16

The sacrificial calendar of the deme Erkhia prescribes sacrifices on this day to Kourotrophos and Artemis Hekate.

Column B, lines 6–13
(Μεταγειτνιῶνος)   (1)
ἕκτηι ἐπὶ δέ-
κα, Κοροτρόφ-
ωι, ἐν ['Ε]κάτης
Ἐρχιᾶσι, χοῖ-
ρος, ⊦⊦⊦
Ἀρτέμιδι Ἑκ-
άτει, Ἐρχιᾶ[σ]-
ι[ν, αἴ]ξ, Δ

Refer also to Metageitnion—Summary.

Restoration to give Metageitnion 16:

A meeting of the Ekklesia in 333/2 B.C., *IG* II² 339a, lines 2–8 as restored by Pritchett and Neugebauer, *Calendars*, pp. 46–48. These lines are restored by Kirchner and Meritt (*Hesp* 1935, p. 532) to give Metageitnion 25.

METAGEITNION

Metageitnion 17

No indisputable evidence as to the nature of this day survives, but refer also to Metageitnion—Summary.

Restorations to give Metageitnion 17:

A meeting of the Ekklesia in 269/8 (?) B.C., *IG* II^2 734, lines 3–7.
A meeting of the Ekklesia in 125/4 B.C., *Hesp* 1933, pp. 163–165, no. 9 (Meritt), lines 1–3 as later revised by Meritt, *Year*, pp. 190–191. Meritt originally restored these lines to give Metageitnion 14.

Metageitnion 18

No indisputable evidence as to the nature of this day survives, but refer also to Metageitnion—Summary.

Restoration to give Metageitnion 18:

A meeting of the Ekklesia in 219/8 B.C., *Hesp* 1960, p. 76, no. 153 (Meritt), lines 1–4.

Metageitnion 19

The sacrificial calendar of the deme Erkhia prescribes a sacrifice to the Heroines on this day.

Column E, lines 1–8
Μεταγειτνι-
ῶνος ἐνάτε
ἐπὶ δέκα, Ἡρ-
ωίναις, ἐπὶ
Σχοίνωι Ἐρ-
χιᾶσι, οἷς, ο-
ὐ φορά, ἱερέ-
αι τὸ δέρ: Δ

Refer also to Metageitnion—Summary.

Restoration to give Metageitnion 19:

A meeting of the Ekklesia in 188/7 B.C., *IG* II^2 891, lines 1–3 as restored by Meritt, *Year*, p. 155. These lines were restored by

III. THE ATHENIAN CALENDAR

Kirchner to give Mounichion 11. Meritt had earlier (*AJP* 78 [1957], p. 381) restored these lines to give Mounichion 19.

Metageitnion 20

The sacrificial calendar of Erkhia designates for this day a sacrifice to Hera Thelkhinia in Erkhia.

Column A, lines 6–11

(Μεταγειτνιῶνος)　　　　　　　　　　(1)
δεκάτει προτ-
έραι, Ἥραι Θελ-
χινίαι, ἐμ Πάγ-
ωι Ἐρχι, ἄρνα π-
αμμέλαιναν, ο-
ὐ φορά, ⌐⊢⊢

For a financial transaction on this day see Metageitnion 20 in Appendix I. No other evidence as to the nature of this day survives, but refer also to Metageitnion—Summary.

Metageitnion 21

Two meetings establish this day as a meeting day for the Ekklesia.

IG II² 796 (see also *Hesp* 1936, p. 203),　　　　　305/4 B.C.
lines 1–4:

[Ἐπ' Εὐξενίππου ἄρχοντος ἐπὶ] τῆς Ἐρεχθεῖδος δευ[τ]-
[έρας πρυτανείας ἧι . . . . ο]ς Λύκου Ἀλωπεκῆθε[ν]
[ἐγραμμάτευεν· Μεταγειτνιῶνο]ς δεκάτει ὑστέραι,
[μιᾶι καὶ εἰκοστῆι τῆς πρυτανεί]ας· ἐκκλησία

The number of the prytany establishes Metageitnion as the month.

IG II² 641, lines 1–7:　　　　　　　　　　　　　　299/8 B.C.

[Ἐπ]ὶ Ε[ὐκτήμον]ος ἄρχοντος ἐπὶ
[τ]ῆς Ἀντιγο[νίδος δ]ευτέρας πρ-
[υ]τανείας, ἐι Θεόφιλος Ξενο[φῶ]-
ντος Κεφαλῆθεν ἐγραμμάτε[υε]-
ν· Μεταγειτνιῶνος δεκάτει ὑστέραι, μιᾶι καὶ εἰκοστεῖ τῆς π-
ρυτανείας· ἐκκλησία

METAGEITNION

Restorations to give Metageitnion 21:

A meeting of the Ekklesia in 175/4 B.C., *Hesp* 1957, pp. 68-71, no. 20 (Meritt), lines 1-3.
A meeting of the Ekklesia in 173/2 B.C., *Hesp* 1957, pp. 33-47, no. 6 (G. A. Stamires), lines 2-4.

Metageitnion 22

No indisputable evidence as to the nature of this day survives.

Restoration to give Metageitnion 22:

A meeting of the Boule in 193/2 B.C., *IG* II^2 864, lines 13-15 as restored by Pritchett and Meritt, *Chronology*, pp. 111-113.

Metageitnion 23

A meeting in 301/0 B.C. establishes this day as a meeting day for the Ekklesia.

IG II^2 640 (see also *Hesp* 1935, p. 547), lines 2-7:

['Επὶ Κλεάρχο]υ ἄρχοντος ἐπὶ τῆς Ἱπποθων-
[τίδος δευτέ]ρας πρυτανείας ἧι Μνήσαρχ-
[ος Τιμοστράτ]ου Προβαλίσιος ἐγραμμάτ-
[ευεν· Μεταγειτ]νιῶνος ὀγδόει μετ' εἰκάδ-
[ας, ἕκτει καὶ εἰκ]οστεῖ τῆς πρυτανείας· ἐκ-
[κλησία]

Metageitnion 24

[Demosthenes] 50.4 provides the only evidence for this day:

Ἑβδόμῃ γὰρ φθίνοντος Μεταγειτνιῶνος μηνὸς ἐπὶ Μόλωνος ἄρχοντος, ἐκκλησίας γενομένης καὶ εἰσαγγελθέντων ὑμῖν πολλῶν καὶ μεγάλων πραγμάτων, ἐψηφίσασθε τὰς ναῦς καθέλκειν τοὺς τριηράρχους.

Molon was, in fact, archon in 362/1 B.C. (Dinsmoor, *Archons*, p. 351), and thus this pseudo-Demosthenic evidence has some basis in fact. For this reason I accept this day as a meeting day for the Ekklesia.

III. THE ATHENIAN CALENDAR

Metageitnion 25

The sacrificial calendar of the deme Erkhia prescribes a sacrifice to Zeus Epopetes on this day.

Column Γ, lines 19-25
([Μ]εταγειτνιῶνος) (13)
ἕκτηι φθίνο-
ντος, Διὶ Ἐπω-
πετεῖ, ἐμ Πάγ-
ωι Ἐρχιᾶσι, χ-
οῖρος, ὀλόκα-
υτος, νηφάλι-
ος, ⊢⊢⊢

For two financial transactions on this day, see Metageitnion 25 in Appendix I.

Restorations to give Metageitnion 25:

A meeting of the Ekklesia in 333/2 B.C., *IG* II² 339a, lines 2-8 (accepted by Meritt, *Hesp* 1935, p. 532). Pritchett and Neugebauer (*Calendars*, pp. 46-48) restore these lines to give Metageitnion 16. A meeting of the Ekklesia in 276/5 B.C., *IG* II² 684, lines 1-5. These lines are restored by Meritt (*Hesp* 1935, pp. 549-550) to give Boedromion 29.

Metageitnion 26

Except for a financial transaction (see Metageitnion 26 in Appendix I), no evidence as to the nature of this day survives.

Metageitnion 27

On this day in 131/0 B.C. there was a meeting of the Boule of the Athenian cleruchs on Salamis.

IG II² 1227, lines 1-2:
Ἐπὶ Ἐπικλέους ἄρχοντος ἐν ἄστει, ἐν Σαλαμῖνι
δὲ Ἀνδρονίκ[ου·]
Μεταγειτνιῶνος τετράδι μετ᾽ εἰκάδας

These Athenian cleruchs surely used the religious and civil calendar of their nearby homeland, and thus the day is established as a meeting day for the Boule.

For a financial transaction on this day, see Metageitnion 27 in Appendix I.

Metageitnion 28

No evidence as to the nature of this day survives.

Metageitnion 29

Three meetings establish this day as a meeting day for the Ekklesia.

Hesp 1932, pp. 45-56 (O. Broneer), 302/1 B.C.
lines 1-5:

['Ε]πὶ Νικοκλέ[ους ἄρχοντος ἐπὶ τῆς Αἰγηί]-
[δ]ος δευτέρα[ς πρυτανείας ἧι Νίκων Θεοδ]-
[ώ]ρου Πλωθεὺ[ς ἐγραμμάτευεν· Μεταγειτν]-
[ι]ῶνος δευτέ[ραι μετ' εἰκάδας, τρίτηι καὶ]
[εἰ]κοστῆι τῆ[ς πρυτανείας· ἐκκλησία]

The number of the prytany establishes that the month must be Metageitnion. [-] κοστῆι (line 5) establishes that the day must be in the last third of the month, and thus only δευτέ [ραι μετ' εἰκάδας] can be restored.

Hesp 1935, pp. 562-565, no. 40 (Meritt), 281/0 B.C.
lines 1-5:

Ἄρχων Οὐρίας· ἐπὶ τῆς Αἰαντίδο[ς δ]ευ[τ]έρ-
ας πρυτανείας ἧι Εὔξενος Καλλί[ου] Αἰξωνεὺς ἐγραμμάτευεν· Μεταγειτνιῶνος δευτέραι μετ' εἰκάδας, ὀγδόει καὶ ε[ἰκο]στῆ-
ι τῆς πρυτανείας· ἐκκλησία

Hesp 1935, pp. 525-530, no. 39 226/5 B.C.
(Meritt), lines 2-7:

Ἐπὶ Ἐργοχάρου ἄρχοντος ἐπὶ τῆς Ἱπποθωντίδος τρίτης πρυτανείας ἧι Ζωίλος Διφί-
λου Ἀλωπεκῆθεν ἐγραμμάτευεν· Μεταγειτνιῶνος δευτέραι μετ' εἰκάδας, ἑβδόμει καὶ εἰ-
κοστῖ τῆς πρυτανείας· ἐκκλησία ἐν τῶι θεάτρωι

III. THE ATHENIAN CALENDAR

Metageitnion 30

Except for a financial transaction (see Metageitnion 30 in Appendix I), no evidence as to the nature of this day survives.

Metageitnion—Summary

The dating of the Eleusinia is the major problem concerning the month Metageitnion. The quadrennial festival occurred in the second year of the Olympiad (*IG* II² 847, lines 9–26), and the biennial festival occurred in the first (*IG* II² 1496, lines 126–130) and third (*IG* II² 1028, lines 4–16 and *IG* II² 2336, lines 200–203) years of the Olympiad.

From *IG* II² 1496, lines 129–133

[ἐκ Πα]ναθηναίων παρὰ ἱερο[ποιῶν: - -]
[ἐξ Ἐλε]υσινίων παρ' ἱεροποιῶ[ν: - -]
[ἐκ τῆς θ]υσί[α]ς τῆι Δημοκρατία[ι παρὰ]
[στρατη]γῶν: [Η]ΗΗΗΔϜϜϜϜΙΙΙ·
[ἐξ Ἀσκλ]ηπιείων παρὰ βοωνῶν: Χ

the Eleusinia occurred after the Panathenaia (Hekatombaion 28) and before the Asklepieia (Boedromion 17 or 18). The Demokratia is tentatively dated to Boedromion 12. Mommsen (*Feste*, pp. 179–190) surveys the activities of the Eleusinia, and assigns four days to them. He then suggests, *exempli gratia*, the dates Boedromion 7–10. But the numerous meetings and festivals in the first half of Boedromion would militate against his dates.

Deubner (*Feste*, p. 91) follows van der Loeff (*De Ludis Eleusiniis*, Leiden, 1903, pp. 79–82) in dating the festival to Metageitnion. The biennial sacrifice to Eleusinia and Kore in Metageitnion in the sacrificial calendar of the Marathonian Tetrapolis (*IG* II² 1358, Column II, lines 43–47) confirms Metageitnion as the month of the Eleusinia. From the study of the days of the month *supra* it may be suggested that the Eleusinia are to be dated to four successive days within the period Metageitnion 13–20.

BOEDROMION

Boedromion 1

This day was a monthly festival day—the Noumenia. For a financial transaction on this day, see Boedromion 1 in Appendix I.

Boedromion 2

This day was a monthly festival day devoted to the Agathos Daimon. This was also the day of the celebration of the Niketeria. The celebration is attested by Proklos *in Ti.* 53D:

ἔτι τοίνυν τῆς Ἀθηνᾶς τὰ νικητήρια παρὰ Ἀθηναίοις ἀνύμνηται, καὶ ἑορτὴν ποιοῦνται ταύτην ὡς τοῦ Ποσειδῶνος ὑπὸ τῆς Ἀθηνᾶς νενικημένου.

The date of the celebration is given by Plutarch in two places.

Mor. 489B:

τὴν γὰρ δευτέραν ἐξαιροῦσιν [Ἀθηναῖοι] ἀεὶ τοῦ Βοηδρομιῶνος, ὡς ἐν ἐκείνῃ τῷ Ποσειδῶνι πρὸς τὴν Ἀθηνᾶν γενομένης τῆς διαφορᾶς.

Mor. 741B:

τὴν δευτέραν τοῦ Βοηδρομιῶνος ἡμέραν ἐξαιροῦμεν, οὐ πρὸς τὴν σελήνην, ἀλλ᾽ ὅτι ταύτῃ δοκοῦσιν ἐρίσαι περὶ τῆς χώρας οἱ θεοί.

Mommsen (*Feste*, p. 171, note 4) rejects the date given by Plutarch on very subjective grounds, "Der 2. ist ein Tag des Haders und Zanks, würdig weggeworfen zu werden—an einem so ungünstigen Tage hat Athena ihr Siegesfest nicht feiern können." Deubner (*Feste*, p. 235, note 2) was sufficiently influenced by Mommsen to question the dating of the festival. I do not believe, however, that there is sufficient reason to contradict Plutarch's explicit statements.

For a financial transaction on this day see Boedromion 2 in Appendix I.

Boedromion 3

This day was a monthly festival day devoted to Athena. And Plutarch, describing the victories which the Athenians ἐσέτι νῦν

47

III. THE ATHENIAN CALENDAR

celebrated with a festival (ἑορτάζειν), also gives this as the day of the victory at Plataea (*Mor.* 349E ff.): τρίτῃ δ' [Βοηδρομιῶνος] ἱσταμένου τὴν ἐν Πλαταιαῖς μάχην ἐνίκων. He gives the same date in *Camillos* 19, Πέρσαι μηνὸς Βοηδρομιῶνος ἕκτῃ μὲν ἐν Μαραθῶνι, τρίτῃ δ' ἐν Πλαταιαῖς ἅμα καὶ περὶ Μυκάλην ἡττήθησαν ὑπὸ τῶν Ἑλλήνων. In *Aristides* 19, however, he gives Boedromion 4 as the date of the battle: ταύτην τὴν μάχην ἐμαχέσαντο τῇ τετράδι τοῦ Βοηδρομιῶνος ἱσταμένου κατ' Ἀθηναίους, κατὰ δὲ Βοιωτοὺς τετράδι τοῦ Πανέμου φθίνοντος, ᾗ καὶ νῦν ἔτι τὸ Ἑλληνικὸν ἐν Πλαταιαῖς ἀθροίζεται συνέδριον καὶ θύουσι τῷ ἐλευθερίῳ Διὶ Πλαταιεῖς ὑπὲρ τῆς νίκης. The celebration he describes here is Plataean rather than Athenian, and *Mor.* 349F supported by *Camillos* 19 would seem to establish Boedromion 3 as the date of the Athenian celebration.[10] There were legal proceedings on this day in the fourth century B.C.

IG II² 1678, lines 27-28:

[ἐμ]ισ[θ]ώ[θ]η Βοηδρομ-
ιῶνος τρίτηι ἱσταμένου δι[κ]αστή[ρ]ιο[ν]

Restoration to give Boedromion 3:

A meeting of the Boule in 210/9-201/0 B.C., *IG* II² 912, lines 35-37 (see also *Hesp* 1957, pp. 59-61, no. 14).

Boedromion 4

This day was a monthly festival day devoted to Herakles, Hermes, Aphrodite, and Eros. On this day in 38/7 B.C. there was a meeting of the Boule.

IG II² 1043, lines 2-4:

Ἀγαθῇ τύχηι τῆς βουλῆς καὶ τοῦ δήμου τοῦ Ἀθηναί[ων·]
[ἐπὶ Κ]αλλικ[ρα]τίδου ἄρ[χο]-
[ντ]ος ἐπὶ τῆς Αἰαντίδος τρίτης πρυτανείας· Βοηδρομ[ιῶνος]
τετράδι ἰστ[αμένου,]
[τ]ετάρτηι τῆς πρυταν[είας· β]ουλὴ ἐν τῶι θεάτρῳ

---

[10] Mommsen, *Feste*, pp. 170-172, and Deubner, *Feste*, p. 235.

BOEDROMION

This meeting, though quite late, is sufficient to establish the day as a meeting day of the Boule. For a financial transaction on this day, see Boedromion 4 in Appendix I.

The sacrificial calendar of Erkhia prescribes a sacrifice to Basile on this day.

Column B, lines 14-20
Βο[ηδ]ρομιῶνο-
ς τετράδι ἱσ-
ταμένο, Βασί-
λει, Ἐρχιᾶ, ἀμ-
νὴ λευκή, ὁλό-
καυτος, νηφά-
λιος, ⌐⊦⊦

Boedromion 5

This was a festival day—the day of the Genesia. The date is established by a lexicographical citation of Philokhoros.

Bekker, *Anecd.* 1.86.20 (Jacoby *FGrHist* 328 F 168):

Γενέσια· οὔσης τε ἑορτῆς δημοτελοῦς Ἀθήναις, Βοηδρομιῶνος πέμπτῃ, Γενέσια καλουμένης, καθότι φησὶ Φιλόχορος καὶ Σόλων ἐν τοῖς ἄξοσι.

The date is confirmed by Dow's (unpublished) recognition of this festival in the fragment of the Athenian State Calendar published as *IG* II^2 1357a, lines 3-22 (see also *Hesp* 1935, p. 23).

The sacrificial calendar of Erkhia prescribes a sacrifice to Epops on this day.

Column Δ, lines 18-23
Βοηδρομιῶνο-
ς πέμπτει ἱσ-
ταμέ: Ἔποπι, Ἐ-
ρχιᾶσι, χοῖρ-
ος, ὁλόκαυτο-
ς, νηφάλι:⊦⊦⊦

Column E, lines 9-15
Βοηδρομιῶν-
ος πέμπτηι
ἱσταμένο, Ἐ-
ρχι: Ἔποπι, χ-
οῖρος, ὁλόκ-
αυτος, νηφά-
λιος,⊦⊦⊦

III. THE ATHENIAN CALENDAR

This wineless holocaust may have some relation to the state festival which is known to have been a festival of the dead.[11]

Restoration to give Boedromion 5:

A meeting of the Ekklesia in 127/6 B.C., *Hesp* 1935, pp. 71-81, no. 37 (Dow), lines 1-3. See also *Hesp* 1955, pp. 220-239. Meritt (*Hesp* 1965, p. 94) restores these lines to give Boedromion 10.

Boedromion 6

The victory at Marathon was celebrated on this day. Plutarch, *Mor.* 349E establishes the date: ἕκτῃ μὲν ἱσταμένου Βοηδρομιῶνος ἐσέτι νῦν τὴν ἐν Μαραθῶνι νίκην ἡ πόλις ἑορτάζει. A procession to Agrai was part of this celebration: Plutarch, *Mor.* 862A, οὐδὲ τὴν πρὸς Ἄγρας πομπὴν ἱστόρηκας, ἣν πέμπουσιν ἔτι νῦν τῇ ἕκτῃ (emended from Ἑκάτῃ) χαριστήρια τῆς νίκης ἑορτάζοντες. This celebration is usually associated with the festival of Artemis Agrotera (Deubner, *Feste*, p. 209) which then must also have been on this day. An annual festival of Artemis on this day would be appropriate, because this day was also the monthly festival day devoted to Artemis.

Two meetings are attested to have occurred on this day.

IG II² 1028, lines 66-68:   101/0 B.C.

Ἀγαθῆι τύχηι· Ἐπὶ Μηδείου ἄρχοντος ἐπὶ τῆς
Λεωντ[ίδο]ς τρίτης πρυτανείας ἧ Φι-
λίων Φιλίωνος Ἐλευσίνιος ἐγ⟦γ⟧ραμμάτευεν·
Βοηδρομιῶνος ἕκτη[ι ἱστα]μένου, ἐνάτῃ τῆς πρυτανεί-
ας· ἐκκλησία κυρία ἐν τῶι θεάτρωι

The scribe erred in recording the day, and the day must be corrected to Boedromion 9. In lines 1-3 the ninth day of the third prytany of the same year is Boedromion 9. The scribe must have mistakenly written ἕκτηι for ἐνάτηι in line 67.

IG II² 1039 (see also Mitsos, *Arkh Eph* 1964, pp. 36-49), lines 1-3:   79/8 B.C.

---

[11] Hesychios Γενέσια, Deubner, *Feste*, pp. 229-230, and Jacoby, *Classical Quarterly* 38 (1944), pp. 65-75.

['Επὶ ..ca.7..] οὐ ἄρχοντος· στρατηγοῦντος ἐπὶ τοὺς
ὁπλίτας Μνασ[έου τοῦ]
Μνασ[έου Βερε]νικίδου· βουλῆς ψηφίσματα· Βοηδρομι[ῶ]νος
ἕκτηι ἱστα[μένου· ἐ]ν τῶι
Θησε[ίωι ἐκ] σταδ[ίου]

There is no clear evidence of scribal error here, and evidently there was a meeting of the Boule on this festival day. This day is also recorded as the appointed day for the handing over of an inventory of property as a result of legal proceedings.

[Demosthenes] 42.1–2:

ἀντὶ μὲν τοῦ τριῶν ἡμερῶν ἀφ' ἧς ὤμοσε τὴν ἀπόφασιν δοῦναί μοι τῆς οὐσίας τῆς αὐτοῦ κατὰ τὸν νόμον, ἢ εἰ μὴ τότ' ἐβούλετο, τῇ γ' ἕκτῃ δοῦναι τοῦ Βοηδρομιῶνος μηνός, ἣν δεηθείς μου ἔθετο καὶ ἐν ᾗ ὡμολόγησε δώσειν τὴν ἀπόφασιν, οὐδέτερα τούτων ἐποίησεν.

Boedromion 7

The seventh day of each month was sacred to Apollo, and for this reason Deubner (*Feste*, p. 202) follows E. Pfuhl (*De Atheniensium Pompis Sacris*, Berlin, 1900, p. 35, note 8) in dating the Boedromia, the festival of Apollo Boedromios, to this day.

Boedromion 8

This was a monthly festival day devoted to Poseidon and Theseus. For financial transactions on this day, see Boedromion 8 in Appendix I.

Restoration to give Boedromion 8:

A meeting of the Ekklesia in 226/5 B.C., *AJP* 63 (1942), p. 422, lines 1–4 (Pritchett) as restored by Meritt, *Year*, p. 154. Pritchett restored these lines to give Anthesterion 8.

Boedromion 9

A meeting in 101/0 B.C. establishes this day as a meeting day for the Ekklesia.

III. THE ATHENIAN CALENDAR

IG II² 1028, lines 1-3:

Ἀγαθῆι τύχηι· Ἐπὶ Μηδείου ἄρχοντος ἐπὶ τῆς
Λεωντίδος τρίτης πρυτανείας ἧ Φιλί-
ων Φιλίωνος Ἐλευσίνιος ἐγραμμάτευεν· Βοηδρομιῶνος
ἐνάτηι ἱσταμένου, ἐνάτῃ
τῆς πρυτανείας· ἐκκλησία κυρία ἐν τῶι θεάτρωι

lines 66-68: see Boedromion 6

Restorations to give Boedromion 9:

A meeting of the Ekklesia in 332/1 B.C., *IG* II² 344, lines 2-9, and *IG* II² 368, lines 1-6.
A meeting of the Ekklesia in 285/4 B.C., *Hesp* 1940, p. 83, no. 14 (Meritt).

Boedromion 10

A meeting in 118/7 B.C. establishes this day as a meeting day for the Ekklesia.

IG II² 1008, lines 1-3:

[Ἐπὶ Ληναίου ἄρχοντ]ος ἐπὶ τῆς Πανδιονίδο[ς] τρίτης
πρυτανείας [ἧ] Ἰσίδωρος Ἀπολ[λωνί]ου Σκαμ[βω]νίδη[ς]
[ἐγραμμάτευεν· Βοη]δρομιῶνος δ[εκά]τηι ἱσταμέν[ου,]
δεκάτῃ τῆς [πρυτα]νείας· ἐκκλησία κ[υ]-
[ρία ἐν τῶι θεάτρωι]

Andocides, 1.121, records legal proceedings on this day: γνοὺς ταῦτα Καλλίας λαγχάνει τῷ υἱεῖ τῷ ἑαυτοῦ τῆς ἐπικλήρου τῇ δεκάτῃ [Βοηδρομιῶνος] ἱσταμένου.

Restorations to give Boedromion 10:

A meeting of the Ekklesia in 219/18 B.C., *Hesp* 1933, pp. 160-161, no. 7 (Meritt), lines 1-5. Meritt later (*Hesp* 1942, pp. 298-299) revised his restoration to give Boedromion 11.
A meeting of the Ekklesia in 127/6 B.C., *Hesp* 1935, pp. 71-81, no. 37 (Dow), lines 1-3 as revised by Meritt in *Hesp* 1965, p. 94. Dow restored these lines to give Boedromion 5. See also *Hesp* 1955, pp. 220-239.

BOEDROMION

Boedromion 11

A meeting in 320/19 B.C. establishes this day as a meeting day for the Ekklesia.

IG II² 380, lines 2–6:

Ἐπὶ Νεαίχμου ἄρχοντος ἐπὶ τῆς Ἐρεχθη-
ίδος δευτέρας πρυτανείας ἕι Θηρα[μ]έν-
ης Κηφισιεὺς ἐγρα[μμ]άτευε· Βοηδρ[ομ]ιῶ-
νος ἐνδεκ[ά]τει, [μ]ιᾶι καὶ τ[ρ]ιακοστἒι τῆ-
ς πρυτ[α]νείας

Phainippos' oath as described in [Demosthenes] 42.11, ὀμόσας γὰρ τῇ ἑνδεκάτῃ τοῦ Βοηδρομιῶνος μηνὸς ἀποφανεῖν ὀρθῶς καὶ δικαίως τὴν οὐσίαν, was clearly made in court, and thus legal proceedings occurred on this day.

Restorations to give Boedromion 11:
A meeting of the Ekklesia in 219/18 B.C., *Hesp* 1933, pp. 160–161 (Meritt), lines 1–5 as revised by Meritt, *Hesp* 1942, pp. 298–299, no. 59. Meritt had earlier restored these lines to give Boedromion 10.

A meeting of the Ekklesia in the second century B.C., *IG* II² 1027, lines 15–16. Pritchett and Meritt (*Chronology*, p. 127) restore these lines to give Pyanopsion 11.

Boedromion 12

According to Plutarch, *Mor.* 349F (τῇ δὲ δωδεκάτῃ [Βοηδρομιῶνος] χαριστήρια ἔθυον ἐλευθερίας· ἐν ἐκείνῃ γὰρ οἱ ἀπὸ Φυλῆς κατῆλθον), the Athenians celebrated the return from Phyle on this day. Deubner (*Feste*, p. 39) follows van der Loeff (*De Ludis Eleusiniis*, pp. 78 ff.) in associating this with the festival named Demokratia in *IG* II² 1496, lines 131–132.

Restoration to give Boedromion 12:
A meeting of the Ekklesia in 244/3 B.C., *Hesp* 1938, pp. 114–115, no. 21 (Meritt), lines 1–4. Meritt later joined this fragment with *IG* II² 766 and has restored the lines to give Mounichion 12 (*Hesp* 1948, pp. 5–7), Pyanopsion 16 (*Χαριστήριον εἰς Ἀναστάσιον Κ. Ὀρλάνδον*, Vol. I, 1965, pp. 193–197), and Posideon 12 (*Year*, p. 148).

III. THE ATHENIAN CALENDAR

Boedromion 13

The activities concerning the Eleusinian Mysteries evidently began on this day. The epheboi went in procession to Eleusis on this day so that they could bring the ἱερά to the Eleusinion in Athens on the following day.

IG II² 1078, lines 9-15:

ἀγαθῆι τύχ[ηι δεδόχθαι] τῶι δήμωι προσ-
τάξαι τῶι κοσμητῆι τῶν [ἐφήβων κ]ατὰ τὰ ἀρχαῖα νόμιμα [ἄ]γειν Ἐλευσῖνάδε τοὺ[ς ἐφήβ]ους τῆι τρίτηι ἐπὶ δέ-
[κα] τοῦ Βοηδρομιῶνος με[τὰ το]ῦ εἰθισμένου σχήμα-
[τος] τῆς ἅμα ἱεροῖς πομπ[ῆς ἵ]να τῆι τετράδι ἐπὶ δέκα πα-
[ραπ]έμψωσιν τὰ ἱερὰ μέχ[ρι] τοῦ Ἐλευσινίου τοῦ ὑπὸ
[τῆι π]όλει.

Although this inscription is quite late (ca. A.D. 220), the proposals are intended to re-establish traditional practices (κατὰ τὰ ἀρχαῖα νόμιμα), and thus we may accept the dates as the traditional dates. There is no indication elsewhere in the sources that this was a state festival day. These processions to and from Eleusis were only preparations for the Mysteries, and there is no need to assume that they were state festivals.

This day is prescribed as the day for a private sacrifice in a calendar of the first or second century A.D.

IG II² 1367, lines 4-6:

Βοηδρομιῶνος γι̅ Νέφθυι καὶ Ὀσίριδ[ι]
ἀλεκτρυόνα καρπώσεις σπείρων πυρο[ὺς]
καὶ κρειθάς, σπένδων μελίκρατον

Boedromion 14

On this day the epheboi escorted the ἱερά from Eleusis to the Eleusinion in Athens (*IG* II² 1078, lines 9-15, cited for Boedromion 13). This was in preparation for the Mysteries, and there is no evidence that this day was a festival day.

A meeting in 106/5 B.C. establishes this day as a meeting day for the Ekklesia.

BOEDROMION

IG II² 1011, lines 2–3:

'Αγαθῆ τύχῃ· Ἐπὶ Ἀγαθοκλέους ἄρχον[τ]ο[ς] ἐπὶ τῆς Αἰαντίδος τρίτης πρυτ[αν]είας ἧ Εὐκλῆς Ξενάνδ[ρου Αἰθαλίδης] ἐγραμμάτευεν· Βοηδρομιῶνος τετράδι ἐ[πὶ] δέκα, τετάρτῃ καὶ δεκάτῃ τῆς π[ρυτ]ανείας· ἐκκλησία κυρία [ἐν τῶι] [θεάτρωι]

For financial transactions on this day, see Boedromion 14 in Appendix I.

Restoration to give Boedromion 14:
A festival of the religious association of the Orgeones in IV B.C., *IG* II² 2501, lines 6–9.

Boedromion 15

This day formed part of the Eleusinian Mysteries and was a festival day—the day of the ἀγυρμός in Athens. The ἀγυρμός is designated by Hesychios (ἀγυρμός· καὶ τῶν μυστηρίων ἡμέρα πρώτη) as the first day of the Mysteries. It therefore must precede the day of ἅλαδε μύσται which is attested to have been Boedromion 16. The procession from Eleusis to Athens occurred on Boedromion 14, and therefore scholars have accepted Boedromion 15 as the day of the ἀγυρμός.[12]

Boedromion 16

Polyaenos 3.11.2 establishes this day as a festival day—the day of ἅλαδε μύσται of the Eleusinian Mysteries.

Χαβρίας περὶ Νάξον ναυμαχῶν ἐνίκησε Βοηδρομιῶνος ἕκτῃ ἐπὶ δέκα, ταύτην τὴν ἡμέραν ἐπιτήδειον τῇ ναυμαχίᾳ κρίνας ὅτι ἦν μία τῶν μυστηρίων. οὕτω γέ τοι καὶ Θεμιστοκλῆς τοῖς Πέρσαις ἐναυμάχησε περὶ Σαλαμῖνα. ἀλλὰ οἱ μὲν περὶ Θεμιστοκλέα σύμμαχον ἔσχον τὸν Ἴακχον, [οἱ δὲ περὶ Χαβρίαν] τὸ ἅλαδε μύσται.

Refer also to Boedromion 16 in Appendix II.

[12] Deubner, *Feste*, p. 72; Dow, *HSCP* 48 (1937), p. 112; G. Mylonas, *Eleusis and the Eleusinian Mysteries*, Princeton, 1961, pp. 247–248.

III. THE ATHENIAN CALENDAR

Boedromion 17

This day was a festival day, but it cannot be determined positively which festival occurred. Deubner (*Feste*, p. 72) argued that this day was the day of the ἱερεῖα δεῦρο of the Eleusinian Mysteries. His argument is that according to Philostratos, *Vita Apoll.* 4.18 (τὰ δὲ Ἐπιδαύρια μετὰ πρόρρησίν τε καὶ ἱερεῖα δεῦρο μυεῖν Ἀθηναίοις πάτριον ἐπὶ θυσίᾳ δευτέρᾳ), the Epidauria occurred after the πρόρρησις (which is associated with the ἀγυρμός on Boedromion 15) and after a day of sacrifice (hence ἐπὶ θυσίᾳ δευτέρᾳ). A private sacrificial calendar from the first or second century A.D. (see *infra*) designates Boedromion 17 as a day for a sacrifice to Demeter and Kore. For this reason Deubner assumes that Boedromion 17 is the first sacrificial day, and that Boedromion 18 is the second sacrificial day. He then naturally dates ἱερεῖα δεῦρο to Boedromion 17 and the Epidauria to Boedromion 18. This dating gives an orderly sequence of one event on each day.[13]

The problems with Deubner's argument are twofold. First, ἐπὶ θυσίᾳ δευτέρᾳ does not necessarily presume two sacrificial days, and, secondly, the sacrifice on the calendar of the first or second century A.D. does not have to be a reflection of the state sacrifice. The evidence would allow the Epidauria to be dated to Boedromion 17 if the ἱερεῖα δεῦρο were dated to Boedromion 16 together with the ἅλαδε μύσται. The Epidauria is, in fact, dated to Boedromion 17 by P. Foucart (*Les Mystères d'Éleusis*, Paris, 1914, pp. 317–323), and Foucart's dating is accepted by P. Graindor (*Athènes sous Hadrien*, Cairo, 1934, p. 153) and is supported by Dow (*HSCP* 48, 1937, p. 113).

In summary, the date of the ἱερεῖα δεῦρο and of the Epidauria cannot be established by the evidence available. The Epidauria may well be dated to either Boedromion 17 or Boedromion 18.

A private sacrifice is prescribed for this day in a calendar from the first or second century A.D.

IG II² 1367, lines 6–7:

(Βοηδρομιῶνος) (4)
ζι Δήμη-
τρι Κόρῃ δέλφακα ἀνυπερθέτως

[13] Mylonas, *Eleusis*, pp. 250–251.

BOEDROMION

Boedromion 18

Five meetings establish this day as a meeting day for the Ekklesia.

IG ii² 657, lines 1-5: 283/2 B.C.

['Ε]πὶ Εὐθίου ἄρχοντος ἐπὶ τῆς 'Α[καμαντίδο]ς τρ[ίτης]
[π]ρυτανείας, ἕι Ναυσιμένης Ναυσικύδου Χολαρ[γεὺ]-
[ς] ἐγραμμάτευεν· Βοιηδρομιῶνος ὀγδόει ἐπὶ δέκ[α, ἐ]-
[ν]άτει καὶ δεκάτει τῆς πρυτανείας· ἐκκλησία κυρ[ί]-
α

Hesp 1936, pp. 418-419, no. 14 (Meritt) 265/4 B.C.
(see also *Hesp* 1969, pp. 110-112),
lines 2-5:

Ἐπὶ Πειθιδή[μ]ου ἄρχ[ον]τος ἐπὶ τῆ[ς 'Α]καμαντίδος τρ[ί]-
[της πρυταν]εί[ας]
[Βοηδρομιῶνος] ὁ[γ]δόει [ἐπ]ὶ δέκα τ[ετάρτε]ι καὶ δεκ[άτει]
[τῆς πρυτανείας· ἐκκλησία]

The restoration to give Boedromion is established by the number of the prytany.

Hesp 1938, pp. 121-123, no. 24 249/8 B.C.
(Meritt), lines 2-4:

['Επὶ Πολυεύ]κτου ἄρχοντος ἐπὶ τῆς Πανδι[ονί]δος τρίτης
  πρυτανείας ἧι
[Χαιρεφῶν 'Αρχ]εσράτου Κεφαλῆθεν ἐγρ[αμμά]τευεν,
  Βοηδρομιῶνος ὀγδόε-
[ι ἐπὶ δέκα, ὀγδόει καὶ] δεκάτηι τῆς πρ[υτανεί]ας·
  ἐκκλησία κυρία

Because the year was ordinary (Pritchett and Neugebauer, *Calendars*, p. 82) the restoration [ἐπὶ δέκα] is required for days 10-19 of the third prytany.

IG ii² 787, lines 1-4: 236/5 B.C.

['Επὶ Ἐκφάντου ἄρχοντο]ς ἐπὶ τῆς 'Αντιο[χίδ]ο[ς] τ[ρ]ίτης
[πρυτανείας ἧι ...⁷...] ος Δημητρίο[υ] Ἱπποτ[ο]μ[ά]δης
[ἐγραμμάτευεν· Βοηδρομι]ῶνος ὀγδόει ἐπὶ δέ[κα,] τετά-
[ρτηι καὶ δεκάτηι τῆς πρυτ]ανείας· ἐκ[κ]λησία

57

III. THE ATHENIAN CALENDAR

The restoration to give Boedromion is established by the number of the prytany.

IG IV² 84, lines 21-24: ca. A.D. 40

Ἐπὶ Σεκούνδου ἄρχοντος καὶ ἱερέως Δρούσου ὑπάτου.
ἐπὶ τῆς Ἐρεχθεῖδος τρίτης πρυτανείας ᾗ Ἀρχέλαος
Λυσιμάχου Μαραθώνιος
ἐγραμμάτευεν· Βοηδρομιῶνος ὀγδόηι ἐπὶ δέκα, ὀγδόηι
καὶ δεκάτηι τῆς πρυτανείας· ἐκκλη-
σία κυρία ἐν τῶι θεάτρωι

The Epidauria may have occurred on this day, but it may equally well have occurred on Boedromion 17 (for discussion see Boedromion 17). If one dates the Epidauria to Boedromion 17, Boedromion 18 would be a regular meeting day and a day of preparation for the procession to Eleusis on Boedromion 19. If one dates the Epidauria to Boedromion 18, then Boedromion 18 must have been a nonfestival day before the Epidauria was inaugurated in Athens in 420 B.C. with the introduction of Asklepios. Meetings of the Ekklesia on Boedromion 18 may have been customary before the introduction of the Epidauria, and for this reason continued to be held despite the celebration of the Epidauria.[14]

A sacrifice is designated for this day on a private sacrificial calendar of the first or second century A.D.

IG II² 1367, lines 7-8:

(Βοηδρομιῶνος)   (4)
ἧι τρυγ[η]-
τὸν Διονύσῳ καὶ τοῖς ἄλλοις θεοῖς ἀνυπερ[θέ](τως)

Restoration to give Boedromion 18:

A meeting of the Ekklesia in 229/8-225/4 B.C., *IG* II² 852, lines 3-6 as restored by Dow, *Hesp* 1963, pp. 364-365.

Boedromion 19

On this day the procession from Athens to Eleusis for the Eleusinian Mysteries occurred.

[14] Deubner, *Feste*, pp. 72-73.

BOEDROMION

IG II² 1078, lines 18–21:

κατὰ τὰ αὐτὰ
[δὲ τῆι] ἐνάτηι ἐπὶ δέκα τοῦ Βοηδρομιῶνος προσ-
[τάξα]ι τῶι κοσμητῆι τῶν ἐφήβων ἄγειν τοὺς ἐφή[βους]
[πάλιν Ἐ]λευσῖνάδε

Several ancient sources date the procession to Boedromion 20: schol. to Aristophanes *Ranae* 324 (μία τῶν μυστηρίων ἐστὶν ἡ εἰκάς, ἐν ᾗ τὸν Ἴακχον ἐξάγουσι); Plutarch *Phoc.* 28 (εἰκάδι γὰρ ἡ φρουρὰ Βοηδρομιῶνος εἰσήχθη μυστηρίων ὄντων, ᾗ τὸν Ἴακχον ἐξ ἄστεος Ἐλευσῖνάδε πέμπουσιν), and *Cam.* 19 (τὴν εἰκάδα τοῦ Βοηδρομιῶνος ᾗ τὸν μυστικὸν Ἴακχον ἐξάγουσιν).

The fact that the procession is dated to two successive days may be explained as follows: the procession left Athens in the morning of Boedromion 19, but did not arrive in Eleusis until after sunset. According to the ancient Greek principle of time reckoning the day began at sunset, and therefore the procession arrived in Eleusis on Boedromion 20.[15] Euripides, *Ion* 1076, places the night festival in Eleusis on Boedromion 20, which would be the evening of the arrival of the procession.

Boedromion 20

No indisputable evidence as to the nature of this day survives. It must have been a festival day of the Eleusinian Mysteries because it was the first full day the μύσται spent at Eleusis. Refer also to Boedromion—Summary. Any reconstruction of the religious activities on this day is highly speculative.

Restoration to give Boedromion 20:

A meeting of the Ekklesia in 245/4 B.C., *IG* II² 799, lines 2–4. Meritt (*Hesp* 1935, p. 550) restores these lines to give Boedromion 30.

Boedromion 21

No evidence as to the nature of this day survives. It was presumably a festival day of the Eleusinian Mysteries—the second full

---
[15] Deubner, *Feste*, p. 72, note 7.

III. THE ATHENIAN CALENDAR

day the μύσται spent at Eleusis. Refer also to Boedromion—Summary.

Boedromion 22

No evidence as to the nature of this day survives, but refer also to Boedromion—Summary.

Boedromion 23

A meeting of a Boule Hiera in Eleusis is attested for this day in A.D. 117/8.

IG II² 1072 as corrected by P. Graindor, *Album d' Inscriptions Attiques d' Époque Impériale*, Ghent, 1924, pp. 28–29, lines 1–3:

Ἐπὶ Τίτου Κωπωνίου, ἱεροκήρυκος υἱοῦ, Μαξίμου
Ἁγνουσίου ἄρχοντος, Βοηδρ[ομιῶνος]
ὀγδόῃ μετ' εἰκάδα, ἐπὶ τῆς Ἀντιοχίδος τρίτης
πρυτανείας, πεντεκαιδεκάτῃ τῆς
πρυτανείας ᾗ Νεικίας Δωρίωνος Φλυεὺς ἐγραμμάτευεν·
βουλὴ ἱερὰ ἐν Ἐλευσεῖνι

Because this was a "sacred" Boule, it may well have met on a festival day. Thus this meeting in Eleusis does not establish the nature of this day as either a festival or meeting day. Refer also to Boedromion—Summary.

Restoration to give Boedromion 23:

A meeting of the Boule (?) in 99/8 B.C. (?), *Hesp* 1963, pp. 23–24, no. 24 (Meritt), lines 2–4.

Boedromion 24

In 222/1 B.C. a meeting of the Boule in the Bouleuterion and in the Eleusinion occurred on this day.

IG II² 848 as re-edited by Dow, *Prytaneis*, pp. 81–85, no. 36, lines 35–39:

BOEDROMION

Ἐπ' Ἀρχελάου ἄρχ[ον]τος [ἐπὶ τῆς] Αἰαντίδος τ[ετ]άρτης πρυτανείας ἧι Μόσχος Μοσ[χίωνος Ἀ]γκυλῆθεν ἐγραμμάτευεν·
Βο-
ηδρομιῶνος ἑβδ[ό]μ[ει μετ' ε]ἰκ[άδα]ς, [τρ]ίτει τῆς πρυτ-
ανείας·
βουλὴ ἐν [βο]υ[λε]υτ[ηρίωι καὶ ἐκ] τοῦ βουλευτηρίου ἐν τῶι Ἐλευ-
σινίωι

This meeting should be associated with the meeting of the Boule in the Eleusinion required by law on the day after the Mysteries.

Andocides 1.111:

ἡ γὰρ βουλὴ ἐκεῖ καθεδεῖσθαι ἔμελλε κατὰ τὸν Σόλωνος νόμον, ὃς κελεύει τῇ ὑστεραίᾳ τῶν μυστηρίων ἕδραν ποιεῖν ἐν τῷ Ἐλευσινίῳ.

Boedromion 25

Two meetings establish this day as a meeting day of the Ekklesia.

IG II² 665, lines 1-4: 282/1 B.C.

[Ἐπ]ὶ Νικίου ἄρχοντος [Ὀτρυνέ]ως ἐπὶ τῆς Ἀκαμαντίδος τρίτ-
[ης] πρυτανείας ἧι Ἰσο[κράτ]ης Ἰσοκράτου Ἀλωπεκῆθεν ἐγρα-
[μμ]άτευεν· Βοηδρομιῶ[νος ἕκτ]ει μετ' εἰκάδας, ἕκτει καὶ εἰκ-
[οσ]τεῖ τῆς πρυτανεία[ς· ἐκκλη]σία

Requirements of spacing guarantee the restoration of [ἕκτ]ει.

IG II² 837, lines 2-6: 227/6 B.C.

[Ἐπὶ] Θεοφίλου ἄρχοντος ἐπὶ τῆς Κεκροπίδ[ος τρίτης]
[πρυτ]ανείας ἧι Φίλιππος Κηφισοδώρου Ἀφ[ιδναῖος]
[ἐγρα]μμάτευεν· Βοηδρομιῶνος ἕκτει μετ' [εἰκάδας,]
[τρίτ]ει καὶ εἰκοστεῖ τῆς πρυτανείας· ἐκ[κλησία ἐν τῶι]
[θεάτρωι]

III. THE ATHENIAN CALENDAR

Refer also to Boedromion 25 in Appendix II.
[Demosthenes] 42.12 (τὴν δὲ ἀπόφασιν τῆς οὐσίας τῇ ἕκτῃ φθίνοντος [Βοηδρομιῶνος]) records that this day was the appointed day to hand over an inventory of property.

Restorations to give Boedromion 25:
A meeting of the Ekklesia in 334/3 B.C., *IG* II^2 335, lines 1–7. These lines are restored by Schweigert (*Hesp* 1940, pp. 339–340) to give Mounichion 25.
A meeting of the Ekklesia in 222/1 B.C., *IG* II^2 848, lines 1–4. Meritt (*Hesp* 1935, p. 557) restores these lines to give Boedromion 27.

Boedromion 26
A meeting of the Areopagos council occurred in Eleusis on this day ca. A.D. 40.

IG IV^2 83, lines 7–8:

Ἐπὶ Σεκούνδου ἄρχοντος καὶ ἱερέως Δρούσου ὑπάτου, μηνὸς
Βοηδρομιῶνος πέμπτῃ ἀπιόντος· Ἄρειος πάγος ἐν
Ἐλευσεῖνι

This assembly may have been a special meeting to treat matters of the Eleusinian Mysteries, and thus does not establish the day as a meeting day.

Boedromion 27
A meeting in 127/6 B.C. establishes this day as a meeting day of the Ekklesia.

Hesp 1935, pp. 71–81, no. 37 (Dow) as re-edited
by Meritt, *Hesp* 1946, pp. 201–213, no. 41,
lines 76–78:

Ἐπὶ Θεοδω[ρίδου ἄρ]χοντος ἐπὶ τῆς Α[ἰγεῖδος τρίτης πρυτ]-
[ανείας ᾗ Σωσικράτης Εὐφρονίου] Θριάσιος ἐγραμμά-
τευεν· Βοη[δρομι]ῶνος τετράδι μετ᾽ εἰκάδας κατ᾽ ἄρχοντα
κατὰ θεὸν δὲ [...⁷... μετ᾽ εἰκάδα]ς, τετάρτῃ καὶ ε[ἰ]-
κοστῇ τῆς [πρυτ]ανείας· ἐκκλη[σία κυρία ἐν τῷ θεάτρῳ]

62

BOEDROMION

The sacrificial calendar of Erkhia prescribes sacrifices to the Nymphs, Akheloos, Alokhos, Hermes, and Ge on this day.

Column A, lines 12–16
Βοηδρομιῶνος
τετράδι φθίν-
οντος, Νύμφαι-
ς, ἐμ Πάγωι Ἐρχ-
ιᾶ, οἷς, Δ

Column B, lines 21–25
(Βο[ηδ]ρομιῶνος)   (14)
τετράδι φθί-
νοντος, ἐμ Πά-
[γ]ωι Ἐρχιᾶσι-
ν, Ἀχελώωι
οἷς, ΔΗ

Column Γ, lines 26–30
[Β]οηδρομιῶνο-
ς τετράδι φθ-
ίνοντος, Ἀλό-
χωι, ἐμ Πάγωι
Ἐρχι: οἷς: Δ

Column Δ, lines 24–27
(Βοηδρομιῶνος)   (18)
τετράδι φθί-
νοντος, Ἑρμῆ-
ι, ἐμ Πάγωι Ἐρ-
χιᾶ: οἷς: ΔΗ

Column E, lines 16–21
(Βοηδρομιῶνος)   (9)
τετράδι φθ-
ίνοντος, Γῆ-
ι, ἐμ Πάγωι Ἐ-
ρχιᾶσι, οἷς
κύουσα, οὐ φ-
ορά, Δ

The sacrificial calendar of the deme Teithras (J. J. Pollitt, *Hesp* 1961, pp. 293–298) prescribes a sacrifice to Athena on this day.

Side A, lines 7–12
(Βοηδ[ρομιῶνος])   (2)
τετράδι φθ[ίνοντος]
ΗΗΗ Ἀθηνᾶι οἶν [- -]
Διί : πρόθυ[μα]
χοῖρον γα[λαθηνόν]
[ἱ]ερεώσυ[να]
[...]εσθη[- - -]

Restoration to give Boedromion 27:

A meeting of the Ekklesia in 222/1 B.C., *IG* II² 848, lines 1–4 as

III. THE ATHENIAN CALENDAR

restored by Meritt, *Hesp* 1935, p. 557. In *IG* II² 848 these lines are restored to give Boedromion 25.

Boedromion 28

No evidence survives as to the nature of this day.

Boedromion 29

No indisputable evidence as to the nature of this day survives.

Restorations to give Boedromion 29:

A meeting of the Ekklesia in 320/19 B.C., *IG* II² 383 b addenda, lines 3–8 as restored by Meritt, *Year*, pp. 113-114. Kirchner restored these lines to give Thargelion 29.

A meeting of the Ekklesia in 276/5 B.C., *IG* II² 684, lines 1–5 as restored by Meritt, *Hesp* 1935, pp. 549–550. These lines were restored by Kirchner to give Metageitnion 25.

A meeting of the Ekklesia ca. 43/2 B.C., *IG* II² 1040, lines 19–21. See also *Hesp* 1965, pp. 255–272.

Boedromion 30

A meeting in 258/7 B.C. establishes this day as a meeting day for the Ekklesia.

IG II² 700 (see also *Hesp* 1938, no. 20, pp. 110–114), lines 1–4:

['E]πὶ Θυμοχάρου ἄρχοντο[ς ἐπὶ τῆς ..... τρίτης πρυτανε]-
[ί]ας ἧι Σώστρατο[ς] 'A[ρι]στ [........¹⁶........ ἐγραμ]-
[μάτευεν· Βοη]-
[δ]ρομιῶνος ἕνει καὶ [νέαι, ...¹⁶........ τῆς πρυτανεία]-
[ς·] ἐκκλησία κυρία

Refer also to Boedromion 30 in Appendix II.

Restorations to give Boedromion 30:

A meeting of the Ekklesia in 335/4 B.C., *IG* II² 331, lines 1–5, and *IG* II² 330, lines 1–4. *IG* II² 331, lines 1–5, is restored by Meritt (*Year*, p. 81) to give Skirophorion 30.

BOEDROMION

A meeting of the Ekklesia in 307/6 B.C., *Hesp* 1933, pp. 398–402, no. 18 (Broneer), lines 2–7.
A meeting of the Ekklesia in 244/3 B.C., *IG* II^2 797, lines 3–6 as restored by Meritt, *Hesp* 1935, p. 555. These lines were restored by Kirchner to give Anthesterion 30.
A meeting of the Ekklesia in 245/4 B.C., *IG* II^2 799, lines 2–4 as restored by Meritt, *Hesp* 1935, p. 500. These lines were restored by Kirchner to give Boedromion 20.
A meeting of the Ekklesia in 195/4 B.C., Pritchett and Meritt, *Chronology*, pp. 110–111, lines 1–4.

Boedromion—Summary

The survey of the days of Boedromion *supra* suggests two probable sequences for the Eleusinian Mysteries:

Boedromion 15	ἀγυρμός	ἀγυρμός
Boedromion 16	ἅλαδε μύσται	ἅλαδε μύσται + ἱερεῖα δεῦρο
Boedromion 17	ἱερεῖα δεῦρο	Epidauria
Boedromion 18	Epidauria + meetings	meetings
Boedromion 19	procession to Eleusis	
Boedromion 20	Mysteries in Eleusis	
Boedromion 21	Mysteries in Eleusis	
Boedromion 22	The Πλημοχόαι[16]	
Boedromion 23	Return to Athens	
	meeting of "sacred" Boule in Eleusis	
Boedromion 24	meeting of Boule in Eleusinion in Athens	

These two sequences have long been considered possible.[17] There is not, however, sufficient evidence to determine which is more probable.

[16] Athenaios 11.496A and Deubner, *Feste*, p. 91.
[17] Dow, *HSCP* 48 (1937), pp. 111–120, and Mylonas, *Eleusis*, pp. 243–280.

PYANOPSION

Pyanopsion 1

This day was a monthly festival day—the Noumenia.

Pyanopsion 2

This day was a monthly festival day devoted to the Agathos Daimon.

Restoration to give Pyanopsion 2:
Legal proceedings in 342/1 (?) B.C., *Hesp* 1936, pp. 393–413, no. 10 (Meritt), lines 11–12.

Pyanopsion 3

This day was a monthly festival day devoted to Athena.

Pyanopsion 4

This was a monthly festival day devoted to Herakles, Hermes, Aphrodite, and Eros.

Pyanopsion 5

Two meetings of private associations are attested for this day: a meeting of Thiasotae in 300/299 B.C.,

IG II² 1263, lines 1–3:

['E]πὶ Ἡγεμάχου ἄρχοντος, μηνὸς Πυανοψιῶνος πέμπτει ἱσταμένου· ἀγορὰ
κυρία τῶν θιασωτῶν

and a meeting of the genos of the Eumolpidae in 152/1 B.C.,

Hesp 1942, pp. 293–298, no. 58 (Meritt), lines 1–3:

Ἐπὶ Λυσιάδου ἄρχοντος, Π[υανοψ]ιῶνος ἕκ[τει ἐπὶ]
δέκα κατὰ θεόν, κατὰ δὲ ἄρ[χοντ]α πέμπτει [ἱστα]-
μένου· ἀγορᾶι κυρίαι ἐν [......] νδίωι

PYANOPSION

The sacred calendar of Eleusis (*IG* II² 1363 as edited by Dow and R. F. Healey, *Harvard Theological Studies* 21, 1965) establishes this day as the day of the πρόρρησις of the Proerosia (Dow and Healey, pp. 14–17).

lines 3–7:

([Πυανοψιῶνος])
πένπτει ἱσταμένου
ἱεροφάντηι καὶ κήρυκι
ε[ἰ]ς ἄριστον τὴν ἑορτὴν
προαγορεύουσιν τῶν
⊢ΙΙΙ Προηροσίων

There is no evidence that this "announcement" by Eleusinian officials of the forthcoming festival required a festival day either in Athens or in Eleusis.

Pyanopsion 6

This was a monthly festival day devoted to Artemis. Two meetings establish that meetings of the Boule could occur on this day.

Otto Kern, *Die Inschriften von*   209/8 B.C.
Magnesia am Maeander, Berlin, 1900,
no. 37, pp. 27–28, lines 1–4:

[- - - ἄρ]χοντος ἐπὶ τῆ[ς Ἱππ]οθωντίδος πέμ-
[πτης πρυτανεία]ς· Ἀρχικλῆ[ς Χ]αριδήμου Ἐρχιεὺς ἐγραμ-
[μάτευε· βου]λῆς ψήφισμα· Πυανοψιῶνος ἕκτει ἱσταμένου,
[ἑβ]δόμε[ι] τῆς πρυτανείας· βουλεῖ ἐμ βουλευτηρίωι

IG II² 1014, lines 1–5:   109/8 B.C.

Ἐπὶ Ἰάσονος ἄρχοντος τοῦ μετὰ Πολύκλει[τον ἐπὶ τῆς]
[- - - ίδος τετάρτης πρυτανεί]-
ας ᾗ Ἐπιφάνης Ἐπιφάνου Λαμπτρεὺς ἐγρα[μμάτευεν·]
[περὶ ἱερῶν· Καλλικρατίδης (?)]
Καλλικράτου Στειριεὺς γράμματα τάδε π[αρέδωκεν εἰς]
[τὴν βουλὴν καὶ τὸν δῆ]-
μον· Πυανοψιῶνος ἕκτῃ ἱσταμένου, πέ[μπτηι τῆς]
[πρυτανείας· βουλὴ ἐν βουλευτη]-
ρίωι

III. THE ATHENIAN CALENDAR

Two festivals, the Proerosia and the Oschophoria, have been dated to this day. Deubner (*Feste*, p. 68) dated the Proerosia to Pyanopsion 5 on the basis of *IG* II² 1363, lines 3–7 (cited *supra* for Pyanopsion 5), but Dow and Healey (pp. 15–16) have correctly distinguished between the prorresis of the festival and the festival itself:

> The present entry is recorded under the fifth, and editors have generally taken this to be the actual day of the festival, without noticing that the inscription specifies the *prorresis* of the festival rather than the festival itself.... There was the return journey for the Hierophantes and the Herald after what must have been a leisurely *ariston*, and sufficient time had to elapse also for those who wished to attend the festival from the city to proceed to Eleusis, probably in procession. There is no evidence that the sacrifice was an evening or night one, and its connection with the ritual ploughing of the Rharian field seems positively to exclude this. The Proerosia can hardly have taken place on the fifth.
>
> Actually only one day is available. The Pyanopsia, inscribed immediately below the entry for the Proerosia, occurred on the seventh. The Proerosia were therefore on the sixth.

If the date proposed by Dow and Healey is correct, and if the Proerosia were a festival of the Athenian state, the meetings of the Boule on this day would be instances in which a meeting was held on an annual festival day. The evidence suggests, however, that the Proerosia were not a festival of the Athenian state. Similar ritual ploughings occurred in the Piraeus and in the deme Myrrhinos (Dow and Healey, p. 16), and like these the Proerosia, the ritual ploughing of the Rharian field, was probably a festival of only the deme. The Eleusinians, of course, invited the Athenians to attend in the prorresis of Pyanopsion 5, but this invitation does not necessarily make the day of the Proerosia a state festival day.

W. S. Ferguson (*Hesp* 1938, pp. 27–28), followed by Jacoby (*FGrHist* IIIb Suppl., Vol. 2, pp. 215–216, note 137), dates the Oschophoria to this day. The sacred calendar of the genos of the Salaminioi of Heptaphylai and Sounion (Ferguson, *Hesp* 1938, pp. 3–5) prescribes a sacrifice to Theseus on Pyanopsion 6:

line 92:

Πυανοψιῶνος. ἕκτει Θησεῖ ὗν ΔΔΔΔ· εἰς τἆλλα ⊢⊢⊢.

PYANOPSION

Ferguson and Jacoby associate this sacrifice with Plutarch's (*Thes.* 22-23) discussion of the Oschophoria. They then establish the following sequence of festival days: Oschophoria at Phaleron on Pyanopsion 6; Pyanopsia ἐν ἄστει on Pyanopsion 7; Theseia on Pyanopsion 8. The identification of this single sacrifice to Theseus with the Oschophoria is, however, by no means certain. There are numerous religious rites associated with Theseus about this time in Pyanopsion (cf. Plutarch, *Thes.* 4, 22-23, 27, 36), and the particular significance of this one sacrifice to Theseus cannot be ascertained. The Oschophoria certainly did occur on some day about this time, but the precise day cannot be established.

Thus neither the Proerosia nor the Oschophoria are proven instances of festivals occurring on a meeting day. The day of the Oschophoria has not been positively established, and it is by no means certain that the Proerosia were a festival of the Athenian state.

Pyanopsion 7

Harpokration (Πυανόψια· Ἀπολλώνιος καὶ σχεδὸν πάντες οἱ περὶ τῶν Ἀθήνησιν ἑορτῶν γεγραφότες Πυανεψιῶνος ἑβδόμῃ Πυανέψια Ἀπόλλωνι ἄγεσθαί φασι) establishes this as the day of the festival of the Pyanopsia. Suda (Πυανεψιῶνος· Πυανεψιῶνος δὲ ὅτι ἑβδόμῃ τὰ Πυανέψια Ἀπόλλωνι ἄγεσθαί φασι) and Plutarch, *Thes.* 22.4 ([Theseus] θάψας δὲ τὸν πατέρα τῷ Ἀπόλλωνι τὴν εὐχὴν ἀπεδίδου τῇ ἑβδόμῃ τοῦ Πυανεψιῶνος μηνὸς ἱσταμένου· ταύτῃ γὰρ ἀνέβησαν εἰς ἄστυ σωθέντες) confirm Harpokration. The celebration of the Pyanopsia on this date is also recorded in the sacred calendar of Eleusis.

IG II² 1363 (as re-edited by Dow and Healey), lines 8-19:

```
  ( [Πυανοψιῶνος]  )
          ἑβδόμει ἱσταμένου
ΔΔ        Ἀπόλλωνι Πυθίωι (α)ἶξ
          καὶ τὰ ἐφ' ἱεροῖς, προγόνιον
          καὶ τὰ μετὰ τούτου,
          τράπεζαν κοσμῆσαι
          τῶι θεῶι, ἱερεώσυνα ἱερεῖ
     ⟦. . . . ca. 15½ (+ ?) letters erased . . . . ⟧
```

III. THE ATHENIAN CALENDAR

ἱε[ρ]οφάντηι καὶ τα[ῖ]ς
ἱερείαις ταῖς ἐξ Ἐλ[ε]υσῖ[νος]
ἐν τῆι παννυχίδι
παρέχειν σπονδ[άς]
ψαιστὰ κα̣[- -]

A private sacrificial calendar of the first or second century A.D. prescribes a sacrifice to Apollo and Artemis on this day.

IG II^2 1367, lines 9–11:

Πυανοψιῶνος Ἀπόλλωνι καὶ Ἀρτέμιδι ζ π[ό]-
πανον χοινικιαῖον ὀρθόνφαλον καὶ καθήμεν[ον]
δωδεκόνφαλον

The celebration of the Pyanopsia and the private offering listed above support the argument that the seventh day of every month was sacred to Apollo.

Plutarch, *Thes.* 4 (Κοννίδαν ᾧ μέχρι νῦν Ἀθηναῖοι μιᾷ πρότερον ἡμέρᾳ τῶν Θησείων κριὸν ἐναγίζουσι), records a sacrifice to Konnidas on the day before the Theseia. The Theseia are generally dated to Pyanopsion 8, and thus the sacrifice to Konnidas is dated to Pyanopsion 7.

Pyanopsion 8

Plutarch, *Thes.* 36 (θυσίαν δὲ ποιοῦσιν αὐτῷ [Θησεῖ] τὴν μεγίστην ὀγδόῃ Πυανεψιῶνος) establishes this day as the day of the largest sacrifice to Theseus. This is commonly associated with the festival named Theseia,[18] which from *IG* II^2 1496, lines 133–136, must have occurred between the Asklepieia (Boedromion 17 or 18) and the Dionysia in the Piraeus (in Posideon [19]):

[ἐξ Ἀσκλ]ηπιείων παρὰ βοωνῶν: Χ
[ἐχ Θησ]έων παρὰ
[ἱεροποι]ῶν: ΧΗ⊓ΔΔΔ⊢⊢⊢
[ἐγ Διονυσίων τῶν]ἐμ Πει[ραιεῖ]

The activities of the Theseia would seem too numerous to be

[18] Mommsen, *Feste*, pp. 288-289, and Deubner, *Feste*, p. 224.
[19] Deubner, *Feste*, pp. 134–137.

accomplished on one day,[20] but there is no evidence to determine which other day(s) to include.

The offering to the Amazons as recorded by Plutarch, *Thes.* 27 (ἥ τε γινομένη πάλαι θυσία ταῖς Ἀμαζόσι πρὸ τῶν Θησείων), may have occurred on this day or on the previous day.[21] The Kybernesia are often associated with the celebration of the Theseia,[22] and thus are tentatively dated to Pyanopsion 8.

Pyanopsion 9

The scholion to Aristophanes *Thesm.* 834, ἀμφότεραι ἑορταὶ γυναικῶν (Stenia and Skira), τὰ μὲν Στήνια πρὸ δυεῖν τῶν Θεσμοφορίων Πυανεψιῶνος θ̄, establishes this day as the day of the celebration of the Stenia,[23] and thus as a festival day.

Pyanopsion 10

This day is recorded by the scholiast to Aristophanes *Thesm.* 80 (δεκάτῃ [Πυανεψιῶνος] ἐν Ἁλιμοῦντι Θεσμοφόρια ἄγεται) as the day of the celebration of the Thesmophoria in Halimus. The date is confirmed by Photios (Θεσμοφορίων ἡμέραι δ̄· δεκάτῃ Θεσμοφόρια, ἑνδεκάτῃ Κάθοδος, δωδεκάτῃ Νηστεία, τρισκαιδεκάτῃ Καλλιγένεια). If the episode in Plutarch, *Solon* 8.4 ff., refers to this celebration, then the day was certainly a festival day attended by τῶν Ἀθηναίων αἱ πρῶται γυναῖκες. This festival was a later addition to the state celebration of the Thesmophoria.[24]

Pyanopsion 11

The scholiast to Aristophanes *Thesm.* 80 ([Πυανεψιῶνος] ιᾱ γὰρ Ἄνοδος, εἶτα ιβ̄ Νηστεία, εἶτα τρισκαιδεκάτῃ Καλλιγένεια) establishes this day as the day of the Anodos of the state Thesmophoria. This is confirmed by Photios (Θεσμοφορίων, cited *supra*

[20] Mommsen, *Feste*, pp. 292–298.
[21] *Ibid.*, p. 290.
[22] *Ibid.*, p. 290 and Deubner, *Feste*, p. 225, but see also Jacoby, *FGrHist* IIIb Suppl., Vol. 2, pp. 345–346.
[23] For the Stenia see Deubner, *Feste*, pp. 52–53.
[24] See Pyanopsion 12 and Deubner, *Feste*, p. 52.

III. THE ATHENIAN CALENDAR

for Pyanopsion 10)[25] and by Alkiphron 3.39.1-2, who identifies the Anodos as the first day of the state Thesmophoria,

[λανθάνει σε] ἡ νῦν ἑστῶσα σεμνοτάτη τῶν Θεσμοφορίων ἑορτή. ἡ μὲν οὖν Ἄνοδος κατὰ τὴν πρώτην γέγονεν ἡμέραν, ἡ Νηστεία δὲ τὸ τήμερον εἶναι παρ' Ἀθηναίοις ἑορτάζεται, τῇ Καλλιγενείᾳ δὲ εἰς τὴν ἐπιοῦσαν θύουσιν.

A meeting of the Ekklesia occurred on this day in 122/1 B.C.

IG II² 1006, lines 50-51:

Ἐπὶ Νικοδήμου ἄρχοντος ἐπὶ [τ]ῆς Ἀντιοχίδος τετ[άρ]της πρυταν[ε]ίας ἧι Ἐπιγένης Ἐπιγένου Οἰναῖος ἐγραμμάτευεν· Πυαν[οψιῶνος] ἐνδεκάτηι, δεκάτηι τῆς πρυτανείας· ἐκκλη[σία] κυρία ἐν τῶι θε[άτ]ρωι

This is a clear instance of a meeting occurring on a festival day, but perhaps it may be explained by the fact that on this day the women alone engaged in the religious activities.[26] The men may have been free to hold a meeting if they desired.

Restorations to give Pyanopsion 11:

A meeting of the Ekklesia in 171/0 B.C., *Hesp* 1934, pp. 14-18, no. 17 (Meritt), lines 1-5 as read by Pritchett, *Ancient Athenian Calendars on Stone*, University of California Publications in Classical Archaeology, Vol. IV, 1963, pp. 278-279. Meritt reads these lines to give Pyanopsion 30.

A meeting of the Ekklesia in the second century B.C., *IG* II² 1027, lines 15-16 as restored by Pritchett and Meritt, *Chronology*, p. 127. These lines were restored by Kirchner to give Boedromion 11.

Pyanopsion 12

This day was the day of the Nesteia of the state Thesmophoria. The date is established by the scholiast to Aristophanes *Thesm.* 80 (cited for Pyanopsion 11), Photios (Θεσμοφορίων, cited for Pyanopsion 10), and Alkiphron 3.39.1-2 (cited for Pyanopsion 11). The state Thesmophoria were originally a three-day festival, including

[25] For the identification of Kathodos with Anodos see Deubner, *Feste*, p. 59.
[26] Deubner, *Feste*, pp. 54-55.

the Anodos, the Nesteia, and the Kalligeneia. For this reason the day of the Nesteia is occasionally referred to as the Μέση.²⁷

In view of all this collaborating evidence Plutarch, *Dem.* 30 ([Δημοσθένης] κατέστρεψε δ' ἕκτῃ ἐπὶ δέκα τοῦ Πυανεψιῶνος μηνός, ἐν ᾗ τὴν σκυθρωποτάτην τῶν Θεσμοφορίων ἡμέραν ἄγουσαι παρὰ τῇ θεῷ νηστεύουσιν αἱ γυναῖκες) which dates the Nesteia to Pyanopsion 16, must be rejected.

Pyanopsion 13

This was the last day of the state Thesmophoria—the Kalligeneia. The date is established by the scholiast to Aristophanes *Thesm.* 80 (cited for Pyanopsion 11), Photios (Θεσμοφορίων, cited for Pyanopsion 10), and Alkiphron 3.39.1–2 (cited for Pyanopsion 11).

Pyanopsion 14

The sacrificial calendar of the deme Erkhia prescribes a sacrifice on this day to the Heroines.

Column A, lines 17–22
Πυανοψιῶνος τ-
ετράδι ἐπὶ δέ-
κα, Ἡρωίναις, ἐ-
μ Πυλῶνι Ἐρχι,
οἷς, οὐ φορά, ἱε-
ρείαι τὸ δέρ, Δ

No other evidence as to the nature of this day survives.

Pyanopsion 15

No evidence as to the nature of this day survives.

Pyanopsion 16

Three meetings establish this day as a meeting day of the Ekklesia.

²⁷ *Ibid.*, p. 52.

III. THE ATHENIAN CALENDAR

IG VII 4254, lines 2–7:   329/8 B.C.

Ἐπὶ Κηφισοφῶντος ἄρχοντος ἐπὶ τ-
ῆς Ἱπποθωντίδος τρίτης πρυτανείας ἧι Σωστρατίδης Ἐχφάντου Εὐπυρίδης ἐγραμμάτευεν·
ἕκτει ἐπὶ δέκα, τρίτει καὶ τριακοστεῖ τῆς πρυτανείας· ἐκκλησία

The day and number of the prytany require that the month be Pyanopsion.

IG II² 795 (see also *Hesp* 1935, p. 551),   245/4 B.C.
lines 1–4:

Ἐπὶ Θεοφήμου ἄρχοντος ἐπὶ [τῆς - - ίδος τετάρτης]
πρυτανείας ἧι Προκ[λ]ῆς Ἀπ[- - - - - - - ἐγραμ]-
[μά]τευεν· Πυανοψιῶνος ἔκ[τει ἐπὶ δέκα, τετάρτηι καὶ]
[δεκ]άτει τῆς πρυτανε[ίας· ἐκκλησία]

No restoration except [δεκ]άτει will fit in line 4, and this in turn requires that ἔκ[τει ἐπὶ δέκα] be restored in line 3.

IG II² 1011, lines 31–32:   106/5 B.C.

Ἀγα[θῆι] τύχηι· Ἐπὶ Ἀγαθοκλέους ἄρχοντος ἐπὶ τῆς
Κεκροπίδος τετά[ρ]της πρυτανείας ἦ Εὐκλῆς Ξενάνδρου
Αἰθα[λίδη]ς ἐγραμμάτευ-
εν· Πυανοψιῶνος ἕκτῃ ἐπὶ δέκα, πέμπτῃ καὶ δεκάτῃ
τῆς πρυτανείας· ἐκκλησία κυρία ἐν τῷ θεάτ[ρ]ῳ

Plutarch, *Dem.* 30, incorrectly dates the Nesteia of the Thesmophoria to this day (see Pyanopsion 12).

Restorations to give Pyanopsion 16:

A meeting of the Ekklesia in 249/8 B.C., *IG* II² 679, lines 1–4.
A meeting of the Ekklesia in 244/3 B.C., *Hesp* 1938, pp. 114–115, no. 21 + *IG* II² 766, lines 1–4 as restored by Meritt, *Χαριστήριον εἰς Ἀναστάσιον Κ. Ὀρλάνδον*, Vol. I, 1965, pp. 193–197. Meritt had previously restored these lines to give Boedromion 12 (*Hesp* 1938, pp. 114–115, no. 21), Mounichion 12 (*Hesp* 1948, pp. 5–7), and Posideon 12 (*Year*, p. 148).

PYANOPSION

Pyanopsion 17

No evidence as to the nature of this day survives.

Pyanopsion 18

Two meetings establish this day as a meeting day of the Ekklesia.

IG II² 481, lines 1-5: 304/3 B.C.

['Επὶ Φερεκλέους ἄρχ]οντος ἐπὶ τῆς Αἰγεῖδος τετ[ά]-
[ρτης πρυτανείας ἧι 'Επιχ]α[ρῖν]ος Δημοχάρους Γα[ρ]-
[γήττιος ἐγραμμάτευεν· Πυανοψι]ῶνος ὀγδόηι ἐπ[ὶ]
[δέκα, ἐνάτηι καὶ δεκάτηι τῆς πρυτανε]ίας· ἐκκλησ-
[ία κυρία]

The year was ordinary (Pritchett and Neugebauer, *Calendars*, p. 79), and thus the prytany number guarantees the restoration to give Pyanopsion.

Hesp 1947, pp. 170-172, no. 67 116/5 B.C.
(Meritt), lines 2-4:

['Ε]πὶ Σαραπίωνος ἄρχοντος ἐπὶ τῆς Οἰνεῖδος τετάρτης
πρυτανείας ἧι Σο-
φοκλῆς Δημητρίου 'Ιφιστιάδης ἐγραμμάτευεν· Πυανοψιῶνος ὀγδόηι ἐπὶ
δέκα, δεκάτηι τῆς πρυτανείας· ἐκκλησία κυρία ἐν
τῶι θεάτρωι

Restoration to give Pyanopsion 18:

A meeting of the Ekklesia in 323/2 B.C., *IG* II² 367, lines 1-7. Meritt (*Year*, p. 107) restores these lines to give Pyanopsion 19, and Dinsmoor (*Archons*, p. 373) restores them to give Pyanopsion 25.

Pyanopsion 19

No indisputable evidence as to the nature of this day survives, but refer also to Pyanopsion—Summary.

III. THE ATHENIAN CALENDAR

Restoration to give Pyanopsion 19:

A meeting of the Ekklesia in 323/2 B.C., *IG* II^2 367, lines 1-6 as restored by Meritt, *Year*, p. 107. Dinsmoor (*Archons*, p. 373) restores these lines to give Pyanopsion 25. These lines are restored by Kirchner to give Pyanopsion 18.

Pyanopsion 20

No evidence as to the nature of this day survives, but refer also to Pyanopsion—Summary.

Pyanopsion 21

No evidence as to the nature of this day survives, but refer also to Pyanopsion—Summary.

Pyanopsion 22

A meeting in 178/7 B.C. establishes this day as a meeting day for the Ekklesia.

Dow, *Prytaneis*, pp. 120-124, no. 64,
lines 1-3:

['Επὶ Φίλω]νος ἄρχοντος τοῦ μετὰ Μενέδημον ἐπὶ τῆς
 Ἱπποθωντίδος τετάρτης πρυτανείας ἧι Φιλιστί-
[ων Φιλ]ιστίωνος Ποτάμιος ἐγραμμάτευεν· Πυανοψιῶνος
 ἐνάτει μετ' εἰκάδας, τριακοστὲι τῆς πρυ-
[τανε]ίας· ἐκκλησία ἐν τῶι θεάτρωι

Refer also to Pyanopsion 22 in Appendix II.

Pyanopsion 23

No evidence as to the nature of this day survives.

Pyanopsion 24

No evidence as to the nature of this day survives.

PYANOPSION

Pyanopsion 25

Two meetings establish this day as a meeting day for the Ekklesia.

IG II² 769, lines 1-6: 251/0 B.C.

['Επ' 'Αντιμάχου] ἄρχοντος ἐ[πὶ τῆ]ς Αἰ[αν]-
[τίδος τετά]ρτης πρυτανεί[ας ἧι] Χαι[ρι]-
[γένης Χαι]ριγένου Μυρρι[νούσι]ος ἐ[γρ]-
[αμμάτευεν·] Πυανοψιῶνος [ἕκτ]ει μετ' εἰ-
[κάδας, πέμπ]τει καὶ εἰκοστῶι τῆς πρυτ-
[ανείας· ἐκκ]λησία κυρία

The spacing allows only the restoration of [ἕκτ]ει in line 4.

Hesp 1940, pp. 104-111, no. 20 302/1 B.C.
(Pritchett), lines 2-7:

Ἐπὶ Νικοκλέους ἄρχοντος ἐπὶ τῆς
Ἀκαμαντίδος τετάρτης πρυτανείας ἧι Νίκων Θεοδώρου Πλωθεὺς ἐγρ-
αμμάτευεν· Πυανοψιῶνος ἕκτει μ[ε]-
τ' εἰκάδας, πέμπτει καὶ εἰκοστῆι τ-
ῆς πρυτανείας· ἐκκλησία

Restoration to give Pyanopsion 25:

A meeting of the Ekklesia in 323/2 B.C., *IG* II² 367, lines 1-6 as restored by Dinsmoor, *Archons*, p. 373. Meritt (*Year*, p. 107) restores these lines to give Pyanopsion 19. These lines are restored by Kirchner to give Pyanopsion 18.

Pyanopsion 26

No evidence as to the nature of this day survives, but refer also to Pyanopsion—Summary.

Pyanopsion 27

No evidence as to the nature of this day survives, but refer also to Pyanopsion—Summary.

III. THE ATHENIAN CALENDAR

Pyanopsion 28

No evidence as to the nature of this day survives, but refer also to Pyanopsion—Summary.

Pyanopsion 29

A meeting in 273/2 B.C. establishes this day as a meeting day of the Ekklesia.

IG II² 674, lines 1-2:

['Ε]πὶ Γλαυκίππου ἄρχοντος ἐπὶ τῆς 'Αντιοχίδος τετάρτης πρυτανείας ἧι Εὔθ[οινος - -]-
ρίτου Μυρρινούσιος ἐγραμμάτευεν· Πυανοψιῶνος δευτέραι μετ' εἰκά[δας· ἐκκλησία]

Restoration to give Pyanopsion 29:

A meeting of the Ekklesia in 324/3 B.C., *IG* II² 547, lines 1-6. These lines are restored by Pritchett and Meritt (*Chronology*, pp. 2-3) to give Thargelion 29.

Pyanopsion 30

This day was a festival day—the day of the celebration of the Chalkeia. The date is established by Suda (Χαλκεῖα II· ἑορτὴ ἀρχαία καὶ δημώδης πάλαι, ὕστερον δὲ ὑπὸ μόνων ἤγετο τῶν τεχνιτῶν...ἔστι δὲ ἕνη καὶ νέα τοῦ Πυανεψιῶνος), Harpokration (Χαλκεῖα· ἑορτὴ παρ' 'Αθηναίοις ἀγομένη Πυανεψιῶνος ἕνῃ καὶ νέᾳ, χειρώναξι κοινή, μάλιστα δὲ χαλκεῦσιν, ὥς φησιν 'Απολλώνιος ὁ 'Αχαρνεύς), and *Etym. Magn.* (805.43, Χαλκεῖα·...ἔστι δὲ ἕνη καὶ νέα τοῦ Πυανεψιῶνος). The date is confirmed by Eustathios on *Il.* 2.552.

Restorations to give Pyanopsion 30:

A meeting of the Ekklesia in 329/8 B.C., *IG* II² 353, lines 1-7.
A meeting of the Ekklesia in 171/0 B.C., *Hesp* 1934, pp. 14-18, no. 17 (Meritt), lines 1-5. See also discussion by Meritt in *Year*, pp. 160-161. Pritchett (*Ancient Athenian Calendars on Stone*, pp. 278-279) reads these lines to give Pyanopsion 11.

PYANOPSION

Pyanopsion—Summary

The exact dates of the major Ionian festival of the Apatouria have never been established. The scholiast to Aristophanes *Ach.* 146 is the best source for the nature and date of the festival:

> λέγει δὲ νῦν περὶ 'Απατουρίων, ἑορτῆς ἐπισήμου δημοτελοῦς, ἀγομένης παρὰ τοῖς 'Αθηναίοις κατὰ τὸν Πυανεψιῶνα μῆνα ἐπὶ τρεῖς ἡμέρας. καλοῦσι δὲ τὴν μὲν πρώτην Δόρπειαν, ἐπειδὴ φράτορες ὀψίας συνελθόντες εὐωχοῦντο· τὴν δὲ δευτέραν 'Ανάρρυσιν, ἀπὸ τοῦ ἀναρρύειν, τοῦ θύειν· ἔθυον δὲ Διὶ Φρατρίῳ καὶ 'Αθηνᾷ· τὴν δὲ τρίτην Κουρεῶτιν, ἀπὸ τοῦ τοὺς κούρους καὶ τὰς κόρας ἐγγράφειν εἰς τὰς φρατρίας.

The Apatouria were thus a three-day festival[28] in Pyanopsion.[29] A fourth day, the Epibda, is occasionally counted as a part of the festival,[30] and Deubner (*Feste*, p. 232, note 5) terms it a "Nachtag" to the festival. From the sacred calendar of the Salaminioi of Heptaphylai and Sounion (*Hesp* 1938, pp. 3–5), lines 92–93,

Πυανοψιῶνος. ἕκτει Θησεῖ ὗν ΔΔΔΔ· εἰς τἄλλα ⊢⊢⊢.
'Απατουρίοις Διὶ Φρατρίωι ὗν ΔΔΔΔ·
ξύλα ἐφ' ἱεροῖς καὶ τἄλλα ⊢⊢⊢. Μαιμακτηριῶνος.
'Αθηνᾶι Σκιράδι οἶν

it is evident that the Apatouria must have occurred after Pyanopsion 6.

The festival required that the citizens assemble in phratries,[31] and thus it would seem highly improbable that public meetings could occur during the three main days of the festival. If this is true, then the Apatouria must be dated to either Pyanopsion 19–21 or Pyanopsion 26–28.

---

[28] See also schol. to Aristophanes *Pax* 890 and *Etym. Magn.* 533.41.
[29] See also Theophrastos *Char.* 3.5 and Harpokration 'Απατούρια.
[30] Harpokration 'Απατούρια, Hesychios 'Απατούρια, and Simplicius to Aristotle *Ph.* 4.11.
[31] Schol. to Aristophanes *Ach.* 146 (*supra*).

MAIMAKTERION

Maimakterion 1

This day was a monthly festival day—the Noumenia.

Maimakterion 2

This day was a monthly festival day devoted to the Agathos Daimon.

Maimakterion 3

This was a monthly festival day devoted to Athena.

Maimakterion 4

This was a monthly festival day devoted to Herakles, Hermes, Aphrodite, and Eros.

Restoration to give Maimakterion 4:

A meeting of the Boule in 159/8 B.C., *Hesp* 1946, pp. 140–142, no. 3 (Pritchett), lines 34–38.

Maimakterion 5

A meeting in 166/5 B.C. establishes this day as a meeting day for the Boule.

Hesp 1934, pp. 21–27, no. 19 (Meritt), lines 1–3:

Ἐπὶ ᾿Αχαιοῦ ἄρ[χοντος ἐπὶ τῆς - - - - πέμπτ]ης
πρυτανείας ἧι Ἡρακλέ[ων]
Ναν(ν)άκου Εὐπ[υρίδης ἐγραμμάτευεν· βου]λῆς ψηφίσματα·
Μαιμακτηρ[ιῶ]-
νος πένπτει ἱστα[μένου, ἕκτηι τῆς πρυτανεί]ας·
βουλῆι ἐν βουλευτηρίω[ι]

Maimakterion 6

This day was a monthly festival day devoted to Artemis. A meeting in 178/7 B.C. also establishes this day as a meeting day for the Boule.

Dow, *Prytaneis*, pp. 120-124, no. 64, lines 27-29:

Ἐπὶ [Φ]ίλωνος ἄρχοντος τοῦ [μετ]ὰ Μενέδημον ἐπὶ τῆς
['Ακαμ]αντίδος πέμπτης πρυτανείας ἧι Φιλ[ι]-
[στ]ίων [Φιλ]ιστ[ί]ωνος Ποτάμιο[ς] ἐγραμμάτευεν· Μαιμακτηριῶνος ἕκτει ἱσταμένου, δεκάτει τῆς πρ[υ]-
τ[α]νείας· βουλὴ ἐμ βουλευτηρίωι

Restoration to give Maimakterion 6:

A meeting of the Ekklesia in 118/7 B.C., *Hesp* 1963, pp. 22-23, no. 23 (Meritt), lines 1-4.

Maimakterion 7

This day was a monthly festival day devoted to Apollo.

Maimakterion 8

This day was a monthly festival day devoted to Poseidon and Theseus.

Maimakterion 9

No evidence as to the nature of this day survives.

Maimakterion 10

No evidence as to the nature of this day survives.

Maimakterion 11

A meeting in 319/8 B.C. establishes this day as a meeting day for the Ekklesia.

III. THE ATHENIAN CALENDAR

Hesp 1940, pp. 345-348, no. 44
(Schweigert), lines 4-8:

Ἐπὶ Ἀπολλοδώρου ἄρχοντος ἐπὶ τῆς
[...]ντίδος τετάρτης πρυτανείας· Μ-
[αιμακ]τηριῶνος ἑνδεκάτει, μιᾶι κα-
[ὶ εἰκοσ]τε͂ι τῆς πρυτανείας· ἐκκλησί-
[α κυρία]

Restorations to give Maimakterion 11:
A meeting of the Ekklesia in 334/3 B.C., *IG* II² 336, lines 1-4.
A meeting of the Ekklesia in 333/2 B.C., *IG* II² 340, lines 1-7.

Maimakterion 12
No evidence as to the nature of this day survives.

Maimakterion 13
No evidence as to the nature of this day survives.

Maimakterion 14
No evidence as to the nature of this day survives.

Maimakterion 15
No evidence as to the nature of this day survives.

Maimakterion 16
A meeting in 220/19 B.C. establishes this day as a meeting day for the Boule.

Hesp 1969, pp. 425-431, no. 2 (Traill),
lines 38-41:

Ἐπὶ Μενεκράτου ἄρχοντος ἐπὶ τῆς Πανδιονίδο[ς]
ἕκτης πρυτανείας ἧι Φιλόδρ(ο)μος Σωτάδου Σουνιεὺ[ς]
ἐγραμμάτευεν· Μαιμακτηριῶνος ἕκτει ἐπὶ δέκα, ἕκτ[ει]
καὶ δεκά(τ)ει τῆς πρυτανείας· ἐκ(κ)λησία κυρία·
δήμου ψήφισμ[α]

The scribe, who was generally careless, inscribed ἐκ(κ)λησία κυρία· δήμου ψήφισμ[α] when it was in fact a meeting of the Boule. See line 47 of this inscription and also Traill's commentary, p. 427.

Restorations to give Maimakterion 16:

A meeting of the Ekklesia in 318/7 B.C., *IG* II^2 448, lines 35-38 as restored by Pritchett and Neugebauer, *Calendars*, p. 64. Meritt (*Year*, p. 126) restores these lines to give Maimakterion 25. These lines are restored in *IG* II^2 448 to give Maimakterion 30.

A meeting of the Ekklesia in 221/0 B.C., *IG* II^2 839, lines 6-11 as restored by Meritt, *Year*, p. 174. These lines are restored in *IG* II^2 839 to give Maimakterion 30.

Maimakterion 17

No evidence as to the nature of this day survives.

Maimakterion 18

No evidence as to the nature of this day survives.

Maimakterion 19

No evidence as to the nature of this day survives.

Maimakterion 20

No evidence as to the nature of this day survives.

Maimakterion 21

Two meetings establish this day as a meeting day for the Ekklesia.

IG II^2 669, lines 1-5:   289/8 B.C.

Ἐπὶ Ἀριστωνύμου ἄρχ[οντος ἐπὶ τῆς ....^{9}....]
ος πέμπτης πρυτανεί[ας ἧι Θρασωνίδης (?)Ν ....ο]-
ς Αἰθαλίδης ἐγραμμ[άτευεν· Μαιμακτηριῶνος δ]-
εκάτει ὑστέραι, δε[υτέραι καὶ εἰκοστεῖ τῆς πρ]-
υτανείας· ἐκκλησ[ία κυρία]

III. THE ATHENIAN CALENDAR

The restoration to give Maimakterion is established by the number of the prytany.

Diogenes Laertios 7.10: 260/59 B.C.

Ἐπ' Ἀρρενίδου ἄρχοντος ἐπὶ τῆς Ἀκαμαντίδος πέμπτης πρυτανείας· Μαιμακτηριῶνος δεκάτῃ ὑστέρᾳ, τρίτῃ καὶ εἰκοστῇ τῆς πρυτανείας· ἐκκλησία κυρία

Refer also to Maimakterion 21 in Appendix II.

Restorations to give Maimakterion 21:
    A meeting of the Ekklesia ca. 290–275 B.C., Dow, *Prytaneis*, pp. 38–39, no. 4, lines 1–2.
    A meeting of the Ekklesia in 238/7(?) B.C., *IG* II^2 702, lines 2–5 as re-edited by Dow, *Prytaneis*, pp. 63–64, no. 21. For the possibility of scribal error see Pritchett and Neugebauer, *Calendars*, pp. 72–73.

Maimakterion 22

No evidence as to the nature of this day survives.

Maimakterion 23

No evidence as to the nature of this day survives.

Maimakterion 24

No evidence as to the nature of this day survives.

Maimakterion 25

No indisputable evidence as to the nature of this day survives.

Restoration to give Maimakterion 25:
    A meeting of the Ekklesia in 318/7 B.C., *IG* II^2 448, lines 35–38 as restored by Meritt, *Year*, p. 126. Pritchett and Neugebauer (*Calendars*, p. 64) restore these lines to give Maimakterion 16. These lines are restored in *IG* II^2 448 to give Maimakterion 30.

MAIMAKTERION

Maimakterion 26
No evidence as to the nature of this day survives.

Maimakterion 27
No indisputable evidence as to the nature of this day survives.

Restoration to give Maimakterion 27:
A meeting of the Ekklesia in 336/5 B.C., *IG* II^2 328, lines 1–6.

Maimakterion 28
No evidence as to the nature of this day survives.

Maimakterion 29
A meeting in 140/39 B.C. establishes this day as a meeting day for the Boule.

Hesp 1948, pp. 17–22, no. 9 (Meritt), lines 36–39:

['Ε]πὶ Ἁγνοθέου ἄρχοντος ἐπὶ τῆς Ἀτταλίδος πέμπτης πρυτανείας
[ἧ]ι Μενεκράτης Χαριξένου Θορίκιος ἐγραμμάτευεν·
Μαιμακτηριῶνος
δευτέραι μετ' εἰκάδας, ὀγδόει καὶ δεκάτει τῆς πρυτανείας· βουλὴ
[ἐ]μ Πειραιεῖ ἐν τῶι Φωσφορίωι

Restoration to give Maimakterion 29:
A meeting of the Ekklesia in 327/6 B.C., *IG* II^2 356, lines 1–8. These lines are restored by Meritt (*Year*, pp. 98–99) to give Anthesterion 29.

Maimakterion 30
A meeting in 321/0 B.C. establishes this day as a meeting day for the Ekklesia.

III. THE ATHENIAN CALENDAR

Hesp 1961, pp. 289-292, no. 84
(Meritt), lines 3-7:

['Επὶ 'Αρχίππ]ου ἄρχοντος ἐπὶ τῆς Λ[ε]-
[ωντίδος πέμπτης π]ρυτανείας ἧι Σωκρατ[.]
[......$\overset{14}{....}$......] δης ἐγραμμάτευεν· Μαι[μ]-
[ακτηριῶνος ἕνει] καὶ νέαι, πέμπτει τῆς [πρ]-
[υτανείας· ἐκκλησ]ία

Restorations to give Maimakterion 30:

A meeting of the Ekklesia in 318/7 B.C., *IG* II^2 448, lines 35-38. These lines are restored by Meritt (*Year*, p. 126) to give Maimakterion 25, and by Pritchett and Neugebauer (*Calendars*, p. 64) to give Maimakterion 16.

A meeting of the Ekklesia in 307/6 B.C., *IG* II^2 456, lines 1-5 as restored by Meritt, *Hesp* 1935, p. 538.

A meeting of the Ekklesia in 221/0 B.C., *IG* II^2 839, lines 6-11. These lines are restored by Meritt (*Year*, p. 174) to give Maimakterion 16.

Maimakterion—Summary

The fact that only one annual religious festival[32] is attested to have occurred during this month is remarkable. This is certainly not an accident of preservation, because Maimakterion is the only month not designated for sacrifices on the sacred calendar of the deme Erkhia. Maimakterion came after the ploughing and planting season[33] and before the harvest season. This may explain the lack of agricultural festivals. Maimakterion fell in early winter, and therefore there are no festivals in commemoration of military victories. The cold and damp weather may also explain why relatively few public meetings are attested for Maimakterion.

[32] The date for the Pompaia cannot be exactly determined. See Deubner, *Feste*, pp. 157-158.
[33] Plutarch, *Mor.* 378E.

POSIDEON

Posideon 1

This day was a monthly festival day—the Noumenia.

Posideon 2

This day was a monthly festival day devoted to the Agathos Daimon. A meeting in 223/2 B.C. also establishes this day as a meeting day for the Boule.

IG II^2 917 as re-edited with new fragments by Pritchett, *Hesp* 1940, pp. 115–118, no. 23, lines 42–43:

Ποσιδεῶνος δευτέ[ραι ἱσταμένου,] δευτέρ[αι τῆς πρυτανείας·] βουλὴ ἐν βουλευτη[ρίωι]

The spacing and formulae will allow only this restoration.

Restoration to give Posideon 2:

A meeting of the Ekklesia in 263/2 B.C., *IG* II^2 477, lines 1–6 as restored by Meritt, *Hesp* 1938, pp. 141–142. Meritt later (*Hesp* 1969, pp. 435–436) restored these lines to give Posideon 3.

Posideon 3

This day was a monthly festival day devoted to Athena.

Restoration to give Posideon 3:

A meeting of the Ekklesia in 255/4 B.C., *IG* II^2 477, lines 1–6 as restored by Meritt, *Hesp* 1969, pp. 435–436. Meritt earlier (*Hesp* 1938, pp. 141–142) restored these lines to give Posideon 2.

Posideon 4

This day was a monthly festival day devoted to Herakles, Hermes, Aphrodite, and Eros.

III. THE ATHENIAN CALENDAR

Restoration to give Posideon 4:

A meeting of the Ekklesia in 294/3 B.C., *IG* II² 378, lines 1-5 as restored by Pritchett and Neugebauer, *Calendars*, pp. 70-72. Meritt (*Hesp* 1938, pp. 98-100) restores these lines to give Posideon 24. See also Dow, *Hesp* 1963, pp. 345-346.

Posideon 5

A meeting in 220/19 B.C. establishes this day as a meeting day for the Ekklesia.

Hesp 1969, pp. 425-431, no. 2 (Traill), lines 1-4:

Ἐπὶ Μενεκράτου ἄρχοντος ἐπὶ τῆς Οἰνεῖδος ἕκτη-
ς πρυτανείας ἧι Φιλόδρομος Σωτάδου Σουνιεὺς
ἐγραμμάτευεν· Ποσιδεῶνος πένπτει ἱσταμένου,
τετάρτει τῆς πρυτανείας· βουλῆς ψηφίσματα

The scribe, who was generally careless, inscribed βουλῆς ψηφίσματα when it was in fact a meeting of the Ekklesia. See lines 12-13, and also Traill's commentary, p. 427.

The celebration of the Plerosia by the deme Myrrhinos has been dated to this day on the basis of *IG* II² 1183, lines 32-37.[34]

τῆι [δὲ πέμπτ]-
ει θυέτω τὴν πληροσίαν ὁ δήμαρχος τῶ[ι]Διὶ ἀπὸ [᾿ᴾ δραχμῶν]
[κ]-
αἰ νεμέτω τὰ κρέα τἑι ἑβδόμει ἱσταμένου τοῖς π[αροῦσιν κ]-
αἰ συναγοράζουσιν καὶ συνενεχυράζουσιν α [- - - - - - -]
μι·τῆι δὲ ἐνάτει ἐπὶ δέκα τοῦ Ποσιδεῶν[ος] μην[ὸς χρηματίζ]-
[ε]ιν πε[ρὶ Διο]νυσίων, τὰ δὲ ἄλλα πάντα τ[- - - - - - - -]

This *sacra lex* refers to a local celebration of the Plerosia, and does not establish the day as a state festival day.

Posideon 6

This day was a monthly festival day devoted to Artemis.

---

[34] Deubner, *Feste*, p. 68.

POSIDEON

Posideon 7

This day was a monthly festival day devoted to Apollo.

Posideon 8

Deubner (*Feste*, pp. 214-215) dates to this day the Posidea, the festival after which the month is named. Plutarch, *Thes.* 36, καὶ γὰρ Ποσειδῶνα ταῖς ὀγδόαις τιμῶσιν, indicates that Poseidon was commonly honored on the eighth day of the month.[35] Posideon 8 is designated as a day of offering to Poseidon on a private sacrificial calendar of the first or second century A.D.

IG II² 1367, lines 16-18:

Ποσιδεῶνος ἦ ἱσταμένου πόπανον
χοινικιαῖον δωδεκόνφαλον καθήμεγ[ον]
Ποσιδῶνι χαμαιζήλῳ νηφάλιον

Thus it is probable, but by no means certain, that Posideon 8 was the day of the Posidea.

Posideon 9

A meeting in 116/5 B.C. establishes this day as a meeting day of the Ekklesia.

IG II² 1009, lines 28-30:

['Επὶ] Σαραπί[ων]ος ἄρχοντος ἐπὶ τῆς 'Ατταλίδος πέμπτης
πρυτανε[ί]ας ἧι Σοφο-
[κλ]ῆς Δημητ[ρίο]υ 'Ιφιστιάδης ἐγραμμάτευεν· Ποσιδεῶνος
ἐνάτηι ἱσταμένου, [ὀ]γδό-
[ηι κ]αὶ εἰκοστῆι τῆς πρυτανείας· ἐκκλησία ἐν τῶι θεάτρω[ι]

Posideon 10

No evidence as to the nature of this day survives.

Posideon 11

Two meetings establish this day as a meeting day of the Ekklesia.

---

[35] See also Hesychios ὀγδοαῖον.

III. THE ATHENIAN CALENDAR

IG II² 666, lines 1–4:   282/1 B.C.

['Επὶ Νικίου ἄρχοντος] 'Οτρυνέ[ως ἐπὶ τῆς - - δος ἕκτ]-
ης πρυτανείας ἧι 'Ισοκράτης 'Ισοκρά[του 'Αλωπεκῆ]θεν ἐ[γ]-
ραμμάτευεν· Ποσιδεῶνος ἐν[δεκ]άτει, δω[δεκά]τει τῆς π[ρ]-
υτανείας· ἐκκλησία κυρία

IG II² 989 as re-edited with new   104/3 B.C.
fragments by M. L. Lethen, *Hesp* 1957,
pp. 25–28, no. 1, lines 1–4:

['Επὶ 'Ηρακλε]ίδ[ου ἄρχοντος ἐπὶ τῆς Πανδιονίδος ἕκτης]
[πρυτα]-
[νείας ἧι] Θρασύβο[υλος Θεοδότου "Ερμειος ἐγραμμάτευεν·]
[Ποσι]-
δεῶνος ἑνδεκά[τηι, ἑνδεκάτηι τῆς πρυτανείας· ἐκκλησία]
κυρία ἐν τῶι θε[άτρωι]

lines 40–42:

'Επὶ 'Ηρακλείδου ἄρχοντος ἐπὶ τῆς Πανδιονίδος ἕκτης
πρυ[τανείας]
[ἧι] Θρασύβουλος Θεοδότου ["Ε]ρμειος ἐγραμμάτευεν·
Ποσιδεῶν[ος ἑνδε]-
[κ]άτηι, ἑνδεκά[τηι τ]ῆς πρ[υ]τανείας· ἐκκλησία κυρία ἐν
τῶι θε[άτρωι]

Restorations to give Posideon 11:
A meeting of the Ekklesia in 335/4 B.C., *Hesp* 1940, pp. 327–328,
no. 36 (Schweigert), lines 1–5.
A meeting of the Ekklesia in 332/1 B.C., *Hesp* 1936, pp. 413–414,
no. 11 (Meritt), lines 2–6.
A meeting of the Ekklesia in 272/1 B.C., *IG* II² 689, lines 1–4.
See also *Hesp* 1957, p. 55. Kirchner in *IG* II² 689 addenda restored
these lines to give Gamelion 11.

Posideon 12

No indisputable evidence as to the nature of this day survives.

Restorations to give Posideon 12:
A meeting of the Ekklesia in 323/2 B.C., *IG* II² 368, lines 19–23.

These lines are restored by Pritchett and Neugebauer (*Calendars*, p. 57) to give Posideon 29 and by Meritt (*Year*, pp. 107–108) to give an intercalated day.

A meeting of the Ekklesia in 244/3 B.C., *Hesp* 1938, pp. 114–115, no. 21 + *IG* II² 766, lines 1–4 as restored by Meritt, *Year*, p. 148. Meritt has also restored these lines to give Boedromion 12 (*Hesp* 1938, pp. 114–115), Mounichion 12 (*Hesp* 1948, pp. 5–7), and Pyanopsion 16 (Χαριστήριον εἰς ᾿Αναστάσιον Κ. ᾿Ορλάνδον, Vol. I, 1965, pp. 193–197).

Posideon 13

No evidence as to the nature of this day survives.

Posideon 14

No evidence as to the nature of this day survives.

Posideon 15

No evidence as to the nature of this day survives.

Posideon 16

On this day in 352/1 B.C. a group of arbitrators began a series of continuous meetings to settle a boundary dispute.

IG II² 204, lines 10–12:

> τὰ-
> [ς δ' ἕδρας ποεῖν συνεχῶς ἀ]πὸ τῆς ἕκτης ἐπὶ δέκα
> τοῦ Ποσιδεῶ-
> [νος ἕως ἂν διαδικασθῆι] ἐπὶ ᾿Αριστοδήμου ἄρχοντος

The day chosen to begin the series of meetings was opportune, because there were no festivals in the near future to interrupt the meetings.

The sacrificial calendar of Erkhia prescribes sacrifices to Zeus on this day.

III. THE ATHENIAN CALENDAR

Column E, lines 22–30
Ποσιδεῶνος
ἕκτηι ἐπὶ δ-
έκα, Διί, ἐμ Π-
έτρηι Ἐρχι-
ᾶσιν, οἷς, οὐ
φορά, ΔΗΗ
Διὶ Ὁρίωι, Ἐ-
ρχιᾶσι, χοῖ-
ρος, οὐ φο: ΔΗΗ

Professor W. Burkert brought to my attention that "it seems to be no coincidence that the meetings to settle the boundary dispute began on the day of a sacrifice to Zeus Horios."

Restoration to give Posideon 16:

A meeting of the Ekklesia in 323/2 B.C., *IG* II^2 448, lines 1–4. There is dispute concerning the readings. Pritchett and Neugebauer (*Calendars*, pp. 57–59) restore the lines to give Posideon 30. Meritt (*Year*, pp. 108–109) follows Kirchner in restoring the lines to give Posideon 16. Dinsmoor (*Archons*, p. 373) restores these lines to give Posideon 25.

Posideon 17

No evidence as to the nature of this day survives.

Posideon 18

No evidence as to the nature of this day survives.

Posideon 19

A meeting of the Ekklesia in 146/5 B.C. establishes this day as a meeting day of the Ekklesia.

Inscriptions de Délos, no. 1504,
lines 50–52:

'Επὶ 'Επικράτου ἄρχοντος ἐπὶ τῆς Λεωντίδος ἕκτης
πρυτανείας ᾗ Ξε[ν - - -]
Συπαλήττιος ἐγραμμάτευεν· Ποσιδεῶνος ἐνάτῃ ἐπὶ
δέκα, ἐνάτῃ κα[ὶ δεκάτηι τῆς πρυτα]-
[ν]είας· ἐκκλησία ἐν τῶι θεάτρωι

A meeting of the deme Myrrhinos occurred on this day in the fourth century B.C.

IG II² 1183, lines 36–37:

τῆι δὲ ἐνάτει ἐπὶ δέκα τοῦ Ποσιδεῶν[ος] μην[ὸς χρηματίζ]-
[ε]ιν πε[ρὶ Διο]νυσίων

A private sacrificial calendar of the first or second century A.D. designates an offering to the Anemoi for this day.

IG II² 1367, lines 18–20:

(Ποσιδεῶνος)   (16)
                                     θι
'Ανέμοις πόπανον χοινικιαῖον ὀρθό[ν]-
φαλον δωδεκόνφαλον νηφάλιον

Posideon 20

A meeting of the Ekklesia of the Athenian cleruchs on Delos occurred on this day in 147/6 B.C.

Inscriptions de Délos 1501, lines 1–3:

'Επὶ Ἄρχοντος ἄρχοντος, ἐπὶ τῆς 'Ατταλίδος ἕκτης πρυτανείας, Ποσιδεῶνος δεκάτει προτέραι, ἐκκλησία κυρί[α]
ἐν τῶι ἐκκλησιαστηρίωι

This meeting of cleruchs is not sufficient evidence to establish that this was a meeting day in Athens.

Posideon 21

No evidence as to the nature of this day survives.

Posideon 22

No evidence as to the nature of this day survives.

III. THE ATHENIAN CALENDAR

Posideon 23

No evidence as to the nature of this day survives.

Posideon 24

No indisputable evidence as to the nature of this day survives.

Restoration to give Posideon 24:

A meeting of the Ekklesia in 294/3 B.C., *IG* II² 378, lines 1–5 as restored by Meritt, *Hesp* 1938, pp. 98–100. Pritchett and Neugebauer (*Calendars*, pp. 70–72) restore these lines to give Posideon 4. See also Dow, *Hesp* 1963, pp. 345–346.

Posideon 25

A meeting in 188/7 B.C. establishes this day as a meeting day for the Ekklesia.

IG II² 890, lines 1–4:

['Επὶ Συμμά]χου ἄρχοντος ἐπὶ [τῆς - - - ἕκτης-πρυτανε]-
[ία]ς ἧι 'Αρ[χικλ]ῆς Θεοδώρου Θορίκ[ιος ἐγραμμάτευεν·]
[δήμου ψή]-
[φισ]μα· Ποσι[δεῶ]νος [ἕ]κ[τ]ει μετ' εἰκάδα[ς, τρίτηι καὶ]
[δεκάτηι τῆς πρυτ]-
[αν]είας· ἐ[κκλη]σία ἐμ Πειραιεῖ

Restoration to give Posideon 25:

A meeting of the Ekklesia in 323/2 B.C., *IG* II² 343, lines 1–5 as restored by Schweigert, *Hesp* 1940, p. 343. Also *IG* II² 448, lines 1–4 as restored by Dinsmoor, *Archons*, p. 373. *IG* II² 448 was restored by Kirchner to give Posideon 16 and by Pritchett and Neugebauer (*Calendars*, pp. 57–59) to give Posideon 30.

Posideon 26

On this day the festival of the Haloa was celebrated. Photios establishes the date: ῾Αλῷα· Ποσιδεῶνος πέμπτη φθίνοντος· ἑορτή ἐστιν 'Αττική. The month of the celebration is confirmed by Harpokration (Ἁλῷα), who cited Philokhoros as his authority, and

94

by *IG* II² 1672, line 124, which lists expenses for the Haloa in the fifth prytany of 329/8 B.C.

Restorations to give Posideon 26:
The taking of an oath in 167/6 B.C., *IG* II² 951, lines 1–2. Kirchner in the addenda restored these lines to give Thargelion 26.
A meeting of the Ekklesia in 125/4 B.C., *IG* II² 1003, lines 1–3 as restored by Meritt, *Year*, pp. 190–191.

Posideon 27

No evidence as to the nature of this day survives.

Posideon 28

No indisputable evidence as to the nature of this day survives.

Restorations to give Posideon 28:
A meeting of the Ekklesia in 324/3 B.C., *Hesp* 1941, pp. 49–50, no. 12 (Meritt), lines 1–7 as restored by Meritt, *Year*, pp. 104–105.
A meeting of the Ekklesia in 304/3 B.C., *IG* II² 482, lines 1–6.
A meeting of the Ekklesia in 279/8 B.C., *Hesp* 1963, pp. 5–6, no. 6 (Meritt), lines 1–4. Meritt later (*Hesp* 1969, p. 110) restored these lines to give Gamelion 8.

Posideon 29

Two meetings establish this day as a meeting day of the Ekklesia.

Hesp 1934, pp. 6–7, no. 7 (Meritt)   302/1 B.C.
as re-edited with new fragments by
Meritt, *Hesp* 1936, pp. 414–416, no. 12,
lines 2–7:

['Επὶ Νικοκ]λέους ἄρχον[τος ἐπὶ τῆ]-
ς 'Αν[τιγον]ίδος ἑβδόμη[ς πρυτανε]-
ίας ἧ[ι] Ṇ[ί]κων Θεοδώρου [Πλωθεὺς ἐ]-
γραμμά[τ]ευεν· Ποσιδεῶ[νος δευτέ]-
ραι μετ' [εἰ]κάδας, [πρ]ώτ[ηι τῆς πρυτ]-
ανεία[ς· ἐκκλησία ἐν Διονύσου]

III. THE ATHENIAN CALENDAR

IG II² 953, lines 1-4: 160/59 B.C.

['E]πὶ Τυχάνδρου ἄρχοντος ἐπὶ τῆς Ἀκαμ[αντίδος ἕκτης]
[πρυ]-
[τ]ανείας ἧι Σωσιγ[έ]νης Μενεκράτου Μαρ[αθώνιος]
[ἐγραμμάτευ]-
[εν·] Ποσιδεῶνος δευτέραι μετ' εἰκάδας, ἐ[νάτηι]
[τῆς πρυτανεί]-
[ας· ἐ]κκλησία κυρία ἐν τῶι θεάτρωι

Restorations to give Posideon 29:

A meeting of the Ekklesia in 323/2 B.C., *IG* II² 368, lines 19-23 as restored by Pritchett and Neugebauer, *Calendars*, p. 57. Meritt (*Year*, pp. 107-108) restores these lines to give an intercalated day. Kirchner restored these lines to give Posideon 12.

A meeting of the Ekklesia in 167/6 B.C., Dow, *Prytaneis*, pp. 133-135, no. 72, lines 1-4. Meritt (*Year*, p. 183) restores these lines to give an intercalated month.

Posideon 30

No indisputable evidence as to the nature of this day survives.

Restorations to give Posideon 30:

A meeting of the Ekklesia in 327/6 B.C., *IG* II² 357, lines 2-6. These lines are restored by Pritchett and Neugebauer (*Calendars*, p. 53) to give Thargelion 30.

A meeting of the Ekklesia in 323/2 B.C., *IG* II² 448, lines 1-4 as restored by Pritchett and Neugebauer, *Calendars*, pp. 57-59. Meritt (*Year*, pp. 108-109) follows Kirchner in restoring these lines to give Posideon 16. Dinsmoor (*Archons*, p. 373) restored these lines to give Posideon 25.

A meeting of the Boule in A.D. 209/10, *IG* II² 1077, lines 1-5.

Posideon—Summary

Posideon fell in the winter months, and the winter weather may partially explain why relatively few meetings of the Ekklesia are attested for this month.

POSIDEON

The rural Dionysia were celebrated in Posideon.[36] The deme Myrrhinos held a meeting on Posideon 19 to treat matters concerning the celebration of the Dionysia.[37] Deubner[38] follows A. Wilhelm [39] in assuming that this meeting occurred shortly after the celebration of the rural Dionysia in Myrrhinos, and thus the celebration in Myrrhinos may be dated to the days just prior to Posideon 19. It is clear, however, that the rural Dionysia were celebrated on different days in the different demes.[40] This may also partially explain the relatively few meetings of the Ekklesia attested for this month.

The rural Dionysia celebrated in the Piraeus may have been a state festival.[41] This celebration evidently included at least four days.[42] There are, however, so few meetings attested for this month that it is impossible to suggest with confidence possible dates for the Dionysia in the Piraeus.

[36] Theophrastos *Char.* 3.5, schol. to Aeschines 1.43, schol. to Plato *Rep.* 475D.
[37] *IG* II² 1183, lines 36–37, cited for Posideon 19.
[38] Deubner, *Feste*, pp. 134–135.
[39] A. Wilhelm, *Urkunden dramatischer Aufführungen in Athen*, Vienna, 1906, pp. 238 ff.
[40] A. Pickard-Cambridge, *The Dramatic Festivals of Athens*² (revised by John Gould and D. M. Lewis, Oxford, 1968), p. 43.
[41] Deubner, *Feste*, p. 137.
[42] *Hesp* 1946, pp. 206–211, lines 24–26.

GAMELION

Gamelion 1

This day was a monthly festival day—the Noumenia.

Gamelion 2

This day was the monthly festival day of the Agathos Daimon.

Gamelion 3

This day was the monthly festival day devoted to Athena.

Gamelion 4

This day was a monthly festival day devoted to Herakles, Hermes, Aphrodite, and Eros.

Gamelion 5

No evidence as to the nature of this day survives.

Gamelion 6

This day was a monthly festival day devoted to Artemis.

Restoration to give Gamelion 6:

A meeting of the Ekklesia in 320/19 B.C., *Hesp* 1944, pp. 234–241, no. 6 (Meritt), lines 2–6 as restored by Dow, *HSCP* 67 (1963), pp. 67–75. These lines are restored by Meritt (*Hesp* 1944, pp. 234–241 and *Hesp* 1963, pp. 425–432) to give Gamelion 10. Meritt (*Year*, pp. 119–120) also proposed a restoration to give Mounichion 8.

Gamelion 7

The sacrificial calendar of Erkhia prescribes sacrifices to Kourotrophos, Apollo Delphinios, and Apollo Lykeios on this day.

GAMELION

Column A, lines 23-30
Γαμηλιῶνος ἐβ-
δόμηι ἱσταμέ-
νο, Κουροτρόφ-
ωι, ἐν Δελφινί-
ωι Ἐρχ: χοῖρ, ⊢⊢⊢
Ἀπόλλωνι Δελ-
φινίωι, Ἐρχιᾶ,
οἷς, Δ⊢⊢

Column E, lines 31-38
Γαμηλιῶνος
ἑβδόμηι ἱσ-
ταμένο, Ἀπό-
λλωνι Λυκε-
ίωι, Ἐρχιᾶσ-
ι, οἷς, Πυθαι-
σταῖς παρ[α]-
δόσιμος, Δ⊢⊢

These sacrifices confirm the nature of the day—a monthly festival day devoted to Apollo. For a financial transaction on this day, see Gamelion 7 in Appendix I.

Restorations to give Gamelion 7:

A meeting of the Ekklesia in 337/6 B.C., *IG* II² 239, lines 3-7 as restored by Schweigert, *Hesp* 1940, p. 327. Also *Hesp* 1940, pp. 325-327, no. 35 (Schweigert), lines 1-4. *IG* II² 239 was restored by Kirchner to give Gamelion 10.

Gamelion 8

This day was a monthly festival day devoted to Poseidon and Theseus. A meeting in 112/1 B.C. establishes this day also as a meeting day for the Boule.

IG II² 1012, lines 1-7:

Ἐπὶ Διονυσίου ἄρχοντος τοῦ μετὰ
Παράμονον ἐπὶ τῆς Αἰαντίδος ἑ-
βδόμης πρυτανείας ᾗ Λάμιος Τιμού-
χου Ῥαμνούσιος ἐγραμμάτευεν· Γαμηλιῶνος ὀγδόῃ ἱσταμένου, ὀγδό-
ῃ τῆς πρυτανείας· βουλὴ ἐμ βουλευτηρίωι

The sacrificial calendar of the deme Erkhia prescribes sacrifices to Apollo Apotropaios, Apollo Nymphegetes, and the Nymphs on this day.

III. THE ATHENIAN CALENDAR

Column A, lines 31–36   Column Γ, lines 31–37
(Γαμηλιῶνος)   (23)   [Γ]αμηλιῶνος ὀ-
ὀγδόηι ἱσταμ-   γδόηι ἱσταμένου, Ἀπόλλων-   έ, Ἀπόλλωνι Ἀ-
ι Ἀποτροπαίω-   ποτροπαίωι,
ι, Ἐρχιᾶσι, πρὸ-   Ἐρχιᾶσι, αἴξ,
[ς] Παιανιέων,   Πυ[θα]ισταῖς
αἴξ, Δ ⊢⊢   παραδός, Δ ⊢⊢

Column E, lines 39–46
(Γαμηλιῶνος)   (31)
ὀγδόη ἱστα-
μένο, Ἀπόλλ-
ωνι Νυμφη[γ]-
έτει, Ἐρχιᾶ-
σιν, αἴξ, Δ ⊢⊢
Νύμφαις, ἐπ-
ὶ τὂ αὐτοῦ β-
ωμοῦ, αἴξ: Δ

Restoration to give Gamelion 8:

A meeting of the Ekklesia in 279/8 B.C., *Hesp* 1963, pp. 5–6, no. 6 (Meritt), lines 1–4 as restored by Meritt, *Hesp* 1969, p. 110. Meritt earlier (*Hesp* 1963, pp. 5–6) restored these lines to give Posideon 28.

Gamelion 9

A meeting in 282/1 B.C. establishes this day as a meeting day for the Ekklesia.

Hesp 1938, pp. 100–109, no. 18 (Meritt), lines 2–6:

Ἐπὶ Νικίου ἄρχοντος ἐπὶ τῆς Οἰνηίδος
ἑβδόμης πρυτανείας ἧι Θεόφιλος Θεοδότου Ἀχαρνεὺς ἐγραμμάτευεν· Γαμηλιῶνος ἐνάτηι ἱσταμένου, τρίτηι καὶ εἰκοστῆι τῆς πρυτανείας· ἐκκλησία

GAMELION

The sacrificial calendar of the deme Erkhia prescribes a sacrifice to Athena on this day.

Column B, lines 26–31
Γαμηλιῶνος
ἐνάτηι ἰστα-
μένο, Ἡροσου-
ρίοις, ἐμ Πόλ-
ει Ἐρχιᾶσι, Ἀ-
θηνᾶι, ἀμνή, ⌐⊢⊢

Gamelion 10

A meeting of the Ekklesia of the Athenian cleruchs on Delos occurred on this day in 159/8 B.C.

Inscriptions de Délos 1498, lines 1–3:

Ἐπὶ Ἀρισταίχμου ἄρχοντος,
Γαμηλιῶνος δεκάτει ἰσταμένου·
ἐκκλησία κυρία ἐν τῶι ἐκκλησιαστηρίωι

This meeting of cleruchs is not sufficient to establish this day as a meeting day in Athens.

Restorations to give Gamelion 10:

A meeting of the Ekklesia in 337/6 B.C., *IG* II² 239, lines 3-7. Schweigert (*Hesp* 1940, p. 327) restores these lines to give Gamelion 7.

A meeting of the Ekklesia in 320/19 B.C., *Hesp* 1944, pp. 234–241, no. 6 (Meritt), lines 2–6. Dow (*HSCP* 67, 1963, pp. 67–75) restores these lines to give Gamelion 6, but Meritt, after proposing Mounichion 8 (*Year*, pp. 119–120), then defends his original restoration (*Hesp* 1963, pp. 425–432) to give Gamelion 10.

Gamelion 11

Three meetings establish this day as a meeting day for the Ekklesia.

III. THE ATHENIAN CALENDAR

IG II² 450, lines 1-6: 314/3 B.C.

Ἐπὶ Νικοδώρου ἄρχοντος
ἐπὶ τῆς Κεκροπίδος ἕκτη-
ς πρυτανείας· Γαμηλιῶνος
ἑνδεκάτηι, ἕκτηι καὶ εἰκοστῆι τῆς πρυτανείας· ἐκκλησία

IG II² 1011, lines 63-65: 107/6 B.C.

Ἀγαθῇ τύχῃ· Ἐπὶ Ἀριστάρχου ἄρχοντος ἐ[πὶ]
τῆς Αἰαντίδος ἑβδόμης πρυτανείας ᾗ Τελέστης Μ[ηδείου]
[Παια]νιεὺς ἐγραμμάτευεν· Γαμηλιῶνος ἑνδεκάτῃ,
ἑνδεκάτῃ τῆς πρυτανείας· ἐκκλησία κυρία ἐ[ν]
τῶι θεάτρῳ

IG II² 1034, lines 1-3: 99/8 B.C.

[Ἐπὶ Προκλ]έους ἄρχοντος ἐπὶ τῆς Κεκροπίδος ἑβδόμης
πρυτανείας
[ἧι - -]θένης Κλεινίου Κοθωκί[δης ἐ]γραμμάτευεν·
Γαμη[λι]ῶνος ἑνδε[κ]-
[άτηι, ἑ]νδεκάτηι τῆς πρυτανείας· [ἐκ]κλησία κυρία
ἐν [τῶι] θεάτρωι

Restorations to give Gamelion 11:

A meeting of the Ekklesia in 302/1 B.C., *IG* II² 499, lines 1-6.
A meeting of the Ekklesia in 272/1 B.C., *IG* II² 689 addenda, lines 1-4. In *IG* II² 689 these lines were restored to give Posideon 11. See also *Hesp* 1957, p. 55.
A meeting of the Ekklesia ca. 200 B.C., *IG* II² 886, lines 1-4.

Gamelion 12

No evidence as to the nature of this day survives, but refer also to Gamelion—Summary.

Gamelion 13

No evidence as to the nature of this day survives, but refer also to Gamelion—Summary.

Gamelion 14

No evidence as to the nature of this day survives, but refer also to Gamelion—Summary.

Gamelion 15

No evidence as to the nature of this day survives, but refer also to Gamelion—Summary.

Gamelion 16

No indisputable evidence as to the nature of this day survives, but refer also to Gamelion—Summary.

Restorations to give Gamelion 16:

A meeting of the Ekklesia in 328/7 B.C., *IG* II^2 452, lines 1–5. This restoration was originally followed by Meritt (*AJP* 59, 1938, p. 499) and by Pritchett and Neugebauer (*Calendars*, p. 51), but later Meritt (*Year*, pp. 95–96) restored these lines to give Gamelion 18.

A meeting of the Ekklesia in 318/7 B.C., *Hesp* 1939, pp. 30–34, no. 8 (Schweigert), lines 1–7 as restored by Pritchett and Neugebauer (*Calendars*, p. 65). Schweigert restores these lines to give Gamelion 30 and Meritt (*Year*, pp. 126–127) restores them to give Gamelion 18.

A meeting of the Ekklesia in 280/79 B.C., *IG* II^2 670, lines 1–4 as restored by Meritt, *Hesp* 1969, pp. 109–110. These lines were restored first by Kirchner (*IG* II^2 670) to give Elaphebolion 11, and later (*IG* II^2 670 addenda) to give Skirophorion 12. Meritt earlier (*Hesp* 1938, p. 106) had restored these lines to give Skirophorion 11.

Gamelion 17

No evidence as to the nature of this day survives, but refer also to Gamelion—Summary.

III. THE ATHENIAN CALENDAR

Gamelion 18

No indisputable evidence as to the nature of this day survives, but refer also to Gamelion—Summary.

Restorations to give Gamelion 18:

A meeting of the Ekklesia in 328/7 B.C., *IG* II^2 452, lines 1–5 as restored by Meritt, *Year*, pp. 95–96. Earlier Meritt (*AJP* 59, 1938, p. 499) and Pritchett and Neugebauer (*Calendars*, p. 51) followed Kirchner's restoration to give Gamelion 16.

A meeting of the Ekklesia in 318/7 B.C., *Hesp* 1939, pp. 30–34, no. 8 (Schweigert), lines 1–7 as restored by Meritt, *Year*, pp. 126–127. These lines are restored by Schweigert to give Gamelion 30 and by Pritchett and Neugebauer (*Calendars*, p. 65) to give Gamelion 16.

A meeting of the Ekklesia in 310/9 B.C., *IG* II^2 453, lines 2–5 as restored by Pritchett and Neugebauer, *Calendars*, p. 66. Meritt (*Year*, p. 129) accepted Kirchner's restoration to give Gamelion 19.

A meeting of the Ekklesia in 250/49 B.C., *IG* II^2 782, lines 1–4 as restored by B. Leonardos, *Arkh Eph* 1919, p. 75.

Gamelion 19

A private sacrificial calendar of the first or second century A.D. designates this as the day to place ivy on the statues of Dionysos.

IG II^2 1367, line 21:

Γαμηλιῶνος κιττώσεις Διονύσους θ̄ι

Refer also to Gamelion—Summary.

Restoration to give Gamelion 19:

A meeting of the Ekklesia in 310/9 B.C., *IG* II^2 453, lines 2–5. Pritchett and Neugebauer, *Calendars*, p. 66, restore these lines to give Gamelion 18. Meritt (*Year*, p. 129) accepts Kirchner's restoration to give Gamelion 19.

Gamelion 20

No evidence as to the nature of this day survives, but refer also to Gamelion—Summary.

Gamelion 21

No indisputable evidence as to the nature of this day survives, but refer also to Gamelion—Summary.

Restoration to give Gamelion 21:

A meeting of the Ekklesia in 164/3 B.C., Dow, *Prytaneis*, pp. 142–146, no. 79, lines 1–5. See also *Hesp* 1957, pp. 74–77.

Gamelion 22

A meeting in 135/4 B.C. establishes this day as a meeting day for the Boule.

Dow, *Prytaneis*, pp. 112–113, no. 56 as re-edited with new fragments by Pritchett, *Hesp* 1940, pp. 126–133, no. 26, lines 37–39:

Ἐπὶ Διο[νυ]σίου ἄρχοντος τ[οῦ μετὰ Τιμαρχίδην ἐπὶ τῆς]
[Πτ]ολεμαιίδος ὀγδό[ης]
πρυτανε[ί]ας ἧι Θεόλυ[τος Θεοδότου Ἀμφιτροπῆθεν]
[ἐγραμ]μάτευεν· Γαμηλι[ῶνος]
ἐνάτει με[τ' εἰκάδας, τετάρτει τῆς πρυτανείας· βουλῆ]
[ἐμ] βουλευτηρίω[ι]

Gamelion 23

No indisputable evidence as to the nature of this day survives.

Restoration to give Gamelion 23:

A meeting of the Ekklesia in 131/0 B.C., *IG* II² 977, lines 1–3 as restored by Pritchett and Meritt, *Chronology*, p. 131.

Gamelion 24

No evidence as to the nature of this day survives.

Gamelion 25

Two meetings establish this day as a meeting day.

III. THE ATHENIAN CALENDAR

Hesp Suppl. IV, pp. 144–147 181/0 B.C.
(H. A. Thompson), lines 10–11:

> Ἐπὶ Ἱππίου ἄρχοντ[ος] ἐπὶ τῆς Πανδιονίδος ὀ[γδόης]
> [πρυτανείας ἧι Θεοδόσιος Ξενοφαν $\underline{\text{ca. 8}}$ εὖς]
> ἐγραμμάτευεν· Γαμηλιῶνος ἕκτει μετ' εἰκά[δας,]
> [ἐνάτει τῆς πρυτανείας· βουλῆς ψηφίσματα]

IG II² 910 as re-edited by Dow, 169/8 B.C.
Prytaneis, pp. 129–133, no. 71,
lines 1–4:

> ['Ε]πὶ Εὐνίκου ἄρχοντος ἐπὶ τῆς Οἰνεῖδος *υυ[υυυυ]*
> ἑ[β]δόμης π[ρυτα]-
> νείας ἧι Ἱερώνυμος Βοήθου Κηφισιεὺς ἐγραμμάτευεν·
> Γαμηλι[ῶνος]
> [ἕ]κτει μετ' εἰκάδας, δευτέραι [καὶ εἰκοστεῖ] τῆ[ς]
> [πρυτανείας·]
> ἐκκλησία ἐμ Πειραιεῖ

A meeting of the Ekklesia of the Athenian cleruchs on Delos also occurred on this day in 165/4 B.C.

Inscriptions de Délos 1497, lines 1–4:

> Ἐπὶ Πέλοπος ἄρχοντος, Γαμη-
> λιῶνος ἕκτει μετ' εἰκάδας,
> ἐκκλησία κυρία ἐν τῶι ἐκκλη-
> σιαστηρίωι

For private sales on this day see Gamelion 25 in Appendix I. Refer also to Gamelion 25 in Appendix II.

Gamelion 26

No evidence as to the nature of this day survives.

Gamelion 27

A meeting in 206/5 B.C. establishes this day as a meeting day for the Ekklesia.

IG II² 849, lines 1–4:

> ['Επὶ] Καλλιστράτου ἄρχοντος ἐπὶ [τῆς - - - ς ὀγδόης]
> [πρυτα]-

GAMELION

[νεία]ς ἧι ʿΑγνωνίδης ʾΑπατου[ρί]ου [- - - ἐγραμμάτευεν·]
[δήμου ψή]-
[φισμα· Γα]μηλιῶνος τετράδι μ[ετ' εἰκάδας, τρίτηι καὶ]
[εἰκοστῆι (?)   τῆς]
[πρυτανεί]ας· ἐκκλησία ἐμ Πειρ[αιεῖ]

The sacrificial calendar of Erkhia prescribes for this day sacrifices to Kourotrophos, Hera, Zeus Teleios, and Poseidon.

Column Γ, lines 38–41
([Γ]αμηλιῶνος)   (31)
τετράδι φθί-
νοντος, Διὶ Τ-
ελείωι, ἐν Ἥρ-
ας Ἔρχι: οἶς, Δ ⊢⊢

Column Δ, lines 28–32
Γαμηλιῶνος τ-
ετράδι φθίν-
οντος, Ποσει-
δῶνι, ἐν Ἥρας
Ἐρχιᾶ, οἶς, Δ ⊢⊢

Column B, lines 32–39
(Γαμηλιῶνος)   (26)
τετράδι φθί-
νοντος, Κουρ-
οτρόφωι, ἐν Ἥ-
ρας Ἐρχιᾶσι,
χοῖρος, ⊢⊢⊢
Ἥραι, Ἐρχιᾶσ-
ι, οἶς, ἱερέαι
δέρμα, Δ

F. Salviat (*BCH* 88, 1964, pp. 647–649) correctly associates these sacrifices with the celebration of the Theogamia in Athens.

Deubner (*Feste*, p. 177) identified the festival Gamelia, after which the month Gamelion was named, with the Theogamia. Hesychios (Γαμηλιών· ὁ τῶν μηνῶν τῆς Ἥρας ἱερός) affirms that Gamelion was sacred to Hera. Deubner (*Feste*, p. 178) then dated the Theogamia to Gamelion 24 on the basis of Menander, fragment 320 (Kock):

ἐμὲ γὰρ διέτριψεν ὁ
κομψότατος ἀνδρῶν Χαιρεφῶν ἱερὸν γάμον
φάσκων ποιήσειν δευτέρᾳ μετ' εἰκάδας
καθ' αὑτόν, ἵνα τῇ τετράδι δειπνῇ παρ' ἑτέροις·

The μετ' εἰκάδας date should, however, be calculated backward from the end of the month, and thus the Theogamia should, as Salviat noted, be dated to Gamelion 27.

107

III. THE ATHENIAN CALENDAR

Gamelion 28

No indisputable evidence as to the nature of this day survives.

Restoration to give Gamelion 28:
A meeting of the Ekklesia in 137/6 B.C., *IG* II² 974, lines 1-4. See also *Hesp* 1959, pp. 188-194, no. 10. Meritt (*Year*, p. 189) restores these lines to give Thargelion 28.

Gamelion 29

Two meetings establish this day as a meeting day for the Ekklesia.

IG II² 483, lines 1-8: 304/3 B.C.

Ἐπὶ Φερεκλέους ἄρχοντος ἐπ-
ὶ τῆς Οἰνεῖδος ἑβδόμης πρυτανείας ἧι Ἐπιχαρῖνος Δημοχάρους Γαργήττιος ἐγραμμάτ-
ευεν· Γαμηλιῶνος δευτέραι μ-
ετ' εἰκάδας, ἐνάτει καὶ εἰκοστῆι τῆς πρυτανείας· ἐκκλησί-
α

Dow, *Prytaneis*, pp. 112-113, no. 56 as 135/4 B.C.
re-edited with new fragments by
Pritchett, *Hesp* 1940, pp. 126-133,
no. 26, lines 1-4:

Ἐπὶ Διονυσίου ἄρχοντος τοῦ μετὰ Τιμαρχίδην [ἐπὶ]
[τῆς Πτολεμαιίδος]
ὀγδόης πρυτανείας ἧι Θεόλυτος Θεοδότου Ἀμφιτρ[οπῆθεν]
[ἐγραμμά]-
τευεν· Γαμηλιῶνος δευτέραι μετ' εἰκάδας, ἑνδεκ[άτει]
[τῆς πρυτανεί]-
ας· ἐκκλησία κυρία ἐν τῶι θεάτρωι

Gamelion 30

Two meetings establish this day as a meeting day for the Ekklesia.

GAMELION

Hesp 1935, pp. 35-37, no. 5 (Oliver),   318/7 B.C.
lines 1-2:

'Επὶ 'Αρχίππου ἄρχοντος [ἐπὶ τῆς - - - ἑβδόμης πρυτανείας]
[ἧι Θέρσιππος Ἱπποθέρσους 'Αχαρνεὺς ἐγραμμά]-
τευεν· Γαμηλιῶνος ἔνει [καὶ νέαι, ὀγδόηι τῆς]
[πρυτανείας· ἐκκλησία κυρία ἐν Διονύσου]

IG II² 653, lines 1-5:   285/4 B.C.

['Επὶ Δ]ιοτίμου ἄρχοντος ἐπὶ τῆς 'Αντι[γονίδος ἑ]-
[βδό]μης πρυτανείας ἧι Λυσίστρατο[ς 'Αριστομά]-
[χου] Παιανιεὺς ἐγραμμάτευεν· Γα[μηλιῶνος ἔνει]
[καὶ] νέαι, ἐνάτηι καὶ εἰ[κο]στῆι τῆ[ς πρυτανείας·]
[ἐκκ]λησία

Restorations to give Gamelion 30:
A meeting of the Ekklesia in 318/7 B.C., *Hesp* 1939, pp. 30-34, no. 8 (Schweigert), lines 1-7. Pritchett and Neugebauer (*Calendars*, p. 65) restore these lines to give Gamelion 16, and Meritt (*Year*, pp. 126-127) restores them to give Gamelion 18.
A meeting of the Ekklesia in 306/5 B.C., *IG* II² 470, lines 1-6.

Gamelion—Summary

The exact dating of the Lenaia has never been established. Two ancient sources specify Gamelion as the month of the festival:

Schol. to Hesiod *Op.* 504-506:

τῷ Γαμηλιῶνι καθ' ὃν καὶ τὰ Λήναια παρ' 'Αθηναίοις

Bekker, *Anecd.* 1.235:

Διονύσια· ἑορτὴ 'Αθήνησι Διονύσου. ἤγετο δὲ τὰ μὲν κατ' ἀγροὺς μηνὸς Ποσειδεῶνος, τὰ δὲ Λήναια Γαμηλιῶνος, τὰ δὲ ἐν ἄστει 'Ελαφηβολιῶνος.

A youthful Dionysos riding on a ram represents the month Gamelion on the calendar frieze from the church of Hagios Eleutherios (Deubner, *Feste*, p. 123, note 6 and plate 37, figure 16). In view of this evidence, the scholion to Plato *Rep.* 475D (τὰ δὲ Λήναια μηνὸς Μεμακτηριῶνος) must be rejected.

III. THE ATHENIAN CALENDAR

The events of the Lenaia included a procession and dramatic contests,[43] hence at least four days must be assigned to the festival. The parallel of the City Dionysia (see Elaphebolion—Summary) would suggest that more than four days should be assigned to the festival.

The period Gamelion 12–21 is free from meetings, and is sufficient to encompass the festival. The sacrifice to Dionysos Leneus on Mykonos on the twelfth day of Lenaion (the Ionic equivalent of the Attic Gamelion)[44] strongly suggests that Gamelion 12 in Athens was a day of the Lenaia.[45] The "placing of ivy" on the statues of Dionysos designated on a late sacrificial calendar (see Gamelion 19) suggests that the festival may have continued until Gamelion 19.

[43] Aristotle *Ath. Pol.* 57.1 and Demosthenes 21.10.
[44] Dittenberger, *Sylloge*^{3} 1024, line 24.
[45] Mommsen, *Feste*, pp. 374–375.

ANTHESTERION

Anthesterion 1

This day was a monthly festival day—the Noumenia.

Anthesterion 2

This day was a monthly festival day devoted to the Agathos Daimon. The sacrificial calendar of the deme Erkhia prescribes a sacrifice to Dionysos on this day.

Column Γ, lines 42–47

['Α]νθεστηριῶν-
ος δευτέραι
ἱσταμένο, Δι-
ονύσωι, Ἔρχι,
ἔριφος προπ-
τόρθι: Π

Anthesterion 3

This day was a monthly festival day devoted to Athena. A meeting of the Thiasotai of Bendis on Salamis occurred on this day in 244/3 B.C.

SEG 2, no. 9, lines 1–2:

Ἐπὶ Κυδήνορος ἄρχοντος, Ἀνθεστηριῶνος τρίτει
ἱσταμένου
κυρίαι ἀγορᾶι

Anthesterion 4

This day was a monthly festival day devoted to Herakles, Hermes, Aphrodite, and Eros.

Restoration to give Anthesterion 4:

A meeting of the Boule in 164/3 B.C., Dow, *Prytaneis*, pp. 142–146, no. 79, lines 34–38 as re-edited with new fragments by Meritt, *Hesp* 1957, pp. 74–77.

111

III. THE ATHENIAN CALENDAR

Anthesterion 5

No evidence as to the nature of this day survives.

Anthesterion 6

This day was a monthly festival day devoted to Artemis.

Anthesterion 7

This day was a monthly festival day devoted to Apollo.

Anthesterion 8

This was a monthly festival day devoted to Poseidon and Theseus.

Restorations to give Anthesterion 8:
A meeting of the Ekklesia in 303/2 B.C., *IG* II^2 489, lines 1–5.
A meeting of the Ekklesia in 226/5 B.C., *AJP* 63 (1942), p. 422, lines 1–4 (Pritchett). These lines are restored by Meritt (*Year*, p. 154) to give Boedromion 8.

Anthesterion 9

A meeting of the Ekklesia of the Athenian cleruchs on Delos occurred on this day in 147/6 B.C.

Inscriptions de Délos 1502, lines 1–4:

Ἐπ' Ἄρχοντος Ἀθήνησιν ἄρχοντ[ος, ἐπὶ]
[τῆς] Οἰνηίδος [ὀγδό]ης πρυτανείας, Ἀ[νθεσ]-
τηριῶνος [ἐν]άτῃ ἱσταμένου· [ἐκκ]λησία [κυρία]
ἐν τῶι ἐκ[κ]λησιαστηρίωι

This meeting of cleruchs is not sufficient evidence to establish this day as a meeting day in Athens.

Restoration to give Anthesterion 9:
A meeting of the Ekklesia in 318/7 B.C., *IG* II^2 350, lines 1–6 as restored by Schweigert, *Hesp* 1939, p. 33. These lines are restored by Pritchett and Neugebauer (*Calendars*, pp. 65–66) to give

ANTHESTERION

Anthesterion 22 and by Meritt (*Year*, p. 127) to give Anthesterion 24, and by S. Dušaníc (*BCH* 89, 1965, pp. 133-134) to give Anthesterion 25. See also *IG* II² 350 addenda.

Anthesterion 10

No indisputable evidence as to the nature of this day survives.

Restoration to give Anthesterion 10:

A meeting of the Ekklesia in 304/3 B.C., *IG* II² 484, lines 1-5.

Anthesterion 11

This day was a festival day—the first day of the Anthesteria. The month is established by the name of the festival, for which the month is named. The festival lasted three days, Anthesterion 11-13: schol. to Thucydides 2.15.4, ἐπὶ τρεῖς μέ[ν] ἐσ[τι]ν ἑορτὴ ἡμέ[ρας] ῑα, ιβ̄, ῑγ, ἐπίσ[ημός ἐσ]τι δὲ ἡ ιβ̄, [ὡς] καὶ εἶπεν αὐ[τός].

The festival consisted of three parts, the Pithoigia, the Khoes, and the Khytroi: Harpokration (and Suda) Χόες· φησὶ δὲ 'Απολλόδωρος 'Ανθεστήρια μὲν καλεῖσθαι κοινῶς τὴν ὅλην τὴν ἑορτὴν Διονύσῳ ἀγομένην, κατὰ μέρος δὲ Πιθοίγια, Χόας, Χύτρους. Plutarch, *Mor.* 655E (τοῦ νέου οἴνου 'Αθήνησι μὲν ἑνδεκάτῃ μηνὸς ['Ανθεστηριῶνος] κατάρχονται, Πιθοίγια τὴν ἡμέραν καλοῦντες) establishes that this day, Anthesterion 11, was the day of the Pithoigia.

Restoration to give Anthesterion 11:

A meeting of the Ekklesia in 331/0 B.C., *IG* II² 363, lines 1-6 as restored by Meritt, *Year*, pp. 88-89. Meritt earlier (*Hesp* 1941, pp. 47-49) restored these lines to give Anthesterion 16.

Anthesterion 12

This day was a festival day—the second day of the Anthesteria (see Anthesterion 11). Thucydides 2.15.4 would suggest that it was the most important day of the Anthesteria: τὸ τοῦ ἐν λίμναις Διονύσου [ἱερόν], ᾧ τὰ ἀρχαιότερα Διονύσια τῇ δωδεκάτῃ ποιεῖται ἐν μηνὶ 'Ανθεστηριῶνι, ὥσπερ καὶ οἱ ἀπ' 'Αθηναίων Ἴωνες ἔτι καὶ νῦν

III. THE ATHENIAN CALENDAR

νομίζουσιν.[46] Harpokration (and Suda) Χόες· ἑορτή τις ἦν παρ' Ἀθηναίοις ἀγομένη Ἀνθεστηριῶνος δωδεκάτῃ) establishes it as the day of the Khoes (see also Anthesterion 11). On the basis of this evidence the second alternative given by the schol. to Aristophanes *Ach.* 961 (εἰς τοὺς Χόας· ἐπετελεῖτο δὲ Πυανεψιῶνος ὀγδόῃ· οἱ δὲ Ἀνθεστηριῶνος [δω]δεκάτῃ) must be accepted. The first date indicates some confusion with the Theseia which were celebrated on Pyanopsion 8.

Anthesterion 13

This day was a festival day—the third and final day of the Anthesteria (see Anthesterion 11). Harpokration (and Suda), citing Philokhoros, establishes it as the day of the Khytroi: Χύτροι· ἔστι δὲ καὶ Ἀττική τις ἑορτὴ Χύτροι... ἤγετο δὲ ἡ ἑορτὴ Ἀνθεστηριῶνος τρίτῃ ἐπὶ δέκα, ὥς φησι Φιλόχορος ἐν τῷ περὶ ἑορτῶν. The scholion to Aristophanes *Ach.* 1076 (and Suda· Χύτροι) (ἐν μιᾷ ἡμέρᾳ ἄγονται οἵ τε Χύτροι καὶ Χόες ἐν Ἀθήναις... οὕτω Δίδυμος) indicates that the Khoes and Khytroi occurred on the same day. This scholion conflicts with Harpokration (*supra* and also for Anthesterion 12), and must be rejected. Deubner (*Feste*, pp. 99–100) accepts Mommsen's explanation (*Feste*, pp. 384–385) that the Khoes extended past sunset on Anthesterion 12, and thus, with the day beginning at sunset, might be felt to be on the "same day" as the Khytroi of Anthesterion 13. Jacoby (*FGrHist* IIIb Suppl., Vol. 2, pp. 268–270) considers this explanation a "shirking of the problem," and explains the error in the scholion as follows: "Didymos probably said that παρὰ τῷ ποιητῇ (i.e., Aristophanes, *Ach.* 1076) Chytroi and Choes actually or apparently fell on the same day (ἄγονται); he then criticized him quoting Philochoros... and thus proving that the Chytroi, in fact, belonged to the following day."

Anthesterion 14

No indisputable evidence as to the nature of this day survives.

Restoration to give Anthesterion 14:
A meeting of the Ekklesia in 273/2 B.C., *IG* II² 675, lines 1–4.

---

[46] Jacoby (*FGrHist* IIIb Suppl., Vol. 2, p. 268, note 5) assumes that the date, τῇ δωδεκάτῃ, has crept into the text of Thucydides from a marginal comment.

ANTHESTERION

Anthesterion 15

No evidence as to the nature of this day survives.

Anthesterion 16

No indisputable evidence as to the nature of this day survives.

Restoration to give Anthesterion 16:

A meeting of the Ekklesia in 331/0 B.C., *IG* II² 363, lines 1–6 as restored by Meritt, *Hesp* 1941, pp. 47–49. Meritt later (*Year*, pp. 88–89) restored these lines to give Anthesterion 11.

Anthesterion 17

No indisputable evidence as to the nature of this day survives.

Restoration to give Anthesterion 17:

A meeting of the Ekklesia in 303/2 B.C., *IG* II² 490, lines 1–5.

Anthesterion 18

Two meetings establish this day as a meeting day for the Ekklesia.

IG II² 788, lines 1–5:   235/4 B.C.

Ἐπὶ Λυ[σ]ανίου ἄρχοντος ἐπὶ τῆς Οἰνεῖδος ὀγδόης πρυτανείας ἧι Εὔμηλος Ἐμπεδίωνος Εὐωνυμεὺς ἐγραμμάτευεν· Ἀνθεστηριῶνος ὀγδόει ἐ-
πὶ δέκα, ὀγδόηι καὶ δεκάτηι τῆς πρυτανείας· ἐκκλησία κυρία

Hesp 1934, pp. 27–31, no. 20 (Meritt)   163/2 B.C.
(see also *Hesp* 1944, p. 266), lines 2–5:

[Ἐπὶ Ἐρά]στου ἄρχοντος ἐπὶ τῆς Λεωντίδος ὀγδόης
    πρυτανείας ἧι Δη-
[μή]τριο[ς] Ξ[έ]νωνος Ἐπικηφίσιος ἐγραμμάτευεν·
    Ἀνθεστηριῶνο[ς]
[ὀγδ]όη[ι ἐπὶ δέκ]α, ὀγδόηι καὶ δεκάτηι τῆς πρυτανείας·
    ἐκκλησία
[κυρία ἐν τῶι] θ[εά]τρωι

III. THE ATHENIAN CALENDAR

Restorations to give Anthesterion 18:

A meeting of the Ekklesia in 197/6 B.C., *IG* II² 888, lines 1-3.
A meeting of the Ekklesia in 188/7 B.C., *Hesp* 1946, pp. 144-146, no. 6 (Pritchett), lines 1-4 as restored by Meritt, *Year*, p. 156. Pritchett restored these lines to give Skirophorion 18.

Anthesterion 19

Two meetings establish this day as a meeting day for the Ekklesia.

IG II² 651, lines 3-8: 286/5 B.C.

['Επὶ Διοκ]λέους ἄρχοντος ἐπὶ τῆς [.]
[...⁷...]ς ὀγδόης πρυτανείας ἧι[Ξ]-
[ενοφῶν Ν]ικέου ῾Αλαιεὺς ἐγραμμά[τ]-
[ευεν· 'Ανθ]εστηριῶνος ἐνάτηι ἐπὶ δ-
[έκα, ἐνάτ]ηι καὶ δεκάτηι τῆς πρυτα-
[νείας· ἐκ]κλησία

IG II² 978 as restored by Meritt, 189/8 B.C.
Hesp 1957, p. 65, lines 2-7:

['Επὶ Εὐθυκρίτου] ἄρχοντος ἐπὶ τῆς Κεκροπί-
[δος ὀγδόης πρ]υτανείας ἧι Κέφαλος Κεφάλου
[Κυδαντίδ]ης ἐγραμμάτευεν· δήμου ψηφίσ-
[ματα· 'Α]νθεστηριῶνος ἐνάτει ἐπὶ δέκα, μι-
[ᾶι καὶ] τριακοσ[[..]]τῆι τῆς πρυτανείας· ἐκκλη-
[σία ἐμ] Πειραιεῖ

Restoration to give Anthesterion 19:

A meeting of the Ekklesia in 286/5 B.C., *Hesp* 1939, p. 42, no. 10 (Schweigert), lines 1-5.

Anthesterion 20

No indisputable evidence as to the nature of this day survives, but refer also to Anthesterion—Summary.

Restoration to give Anthesterion 20:

A meeting of the Ekklesia in 307/6 B.C., *IG* II² 459, lines 2-4 as restored by Pritchett and Meritt, *Chronology*, pp. 18-19.

ANTHESTERION

Anthesterion 21

No evidence as to the nature of this day survives, but refer also to Anthesterion—Summary.

Anthesterion 22

No indisputable evidence as to the nature of this day survives, but refer also to Anthesterion—Summary.

Restoration to give Anthesterion 22:

A meeting of the Ekklesia in 318/7 B.C., *IG* II² 350, lines 1–6 as restored by Pritchett and Neugebauer, *Calendars*, pp. 65–66. These lines are restored by Schweigert (*Hesp* 1939, p. 33) to give Anthesterion 9 and by Meritt (*Year*, p. 127) to give Anthesterion 24, and by S. Dušanić (*BCH* 89, 1965, pp. 133–134) to give Anthesterion 25. See also *IG* II² 350 addenda.

Anthesterion 23

The scholion to Aristophanes *Nub.* 408 establishes this day as the day of the Diasia, the festival of Zeus Meilikhios: Διασίοισιν· ἑορτὴ Ἀθήνησι Μειλιχίου Διός. ἄγεται δὲ μηνὸς Ἀνθεστηριῶνος ᾗ φθίνοντος. The month is confirmed by the sacrificial calendar of the deme Erkhia, which designates a sacrifice to Zeus Meilikhios at the Diasia in Anthesterion.

Column A, lines 37–43
Ἀνθεστηριῶνο-
ς, Διασίοις, ἐν
ἄστει ἐν Ἄγρας,
Διὶ Μιλιχίωι,
οἷς, νηφάλιος
μέχρι σπλάγχ-
[v]ων, Δ ⊢⊢

See also Anthesterion—Summary.

Anthesterion 24

No indisputable evidence as to the nature of this day survives, but refer also to Anthesterion—Summary.

III. THE ATHENIAN CALENDAR

Restoration to give Anthesterion 24:
A meeting of the Ekklesia in 318/7 B.C., *IG* II^2 350, lines 1-6 as restored by Meritt, *Year*, p. 127. These lines are restored by Schweigert (*Hesp* 1939, p. 33) to give Anthesterion 9 and by Pritchett and Neugebauer (*Calendars*, pp. 65-66) to give Anthesterion 22, and by S. Dušanić (*BCH* 89, 1965, pp. 133-134) to give Anthesterion 25. See also *IG* II^2 350 addenda.

Anthesterion 25
No indisputable evidence as to the nature of this day survives, but refer also to Anthesterion—Summary.

Restoration to give Anthesterion 25:
A meeting of the Ekklesia in 318/7 B.C., *IG* II^2 350, lines 1-6 as restored by S. Dušanić, *BCH* 89 (1965), pp. 133-134. See also *IG* II^2 350 addenda. These lines are restored by Schweigert (*Hesp* 1939, p. 33) to give Anthesterion 9, by Pritchett and Neugebauer (*Calendars*, pp. 65-66) to give Anthesterion 22, and by Meritt (*Year*, p. 127) to give Anthesterion 24.

Anthesterion 26
No evidence as to the nature of this day survives, but refer also to Anthesterion—Summary.

Anthesterion 27
A meeting in 189/8 B.C. establishes this day as a meeting day for the Boule.

Dow, *Prytaneis*, pp. 91-92, no. 41 as
re-edited with a new fragment by Meritt,
Hesp 1957, pp. 63-66, no. 17,
lines 1-4:

['Επὶ Εὐθυκρίτο]υ ἄρχοντος ἐπὶ τῆς Αἰαντίδος ἐνάτης πρυ-
[τανείας ἧι Κέφα]λος Κεφάλου Κυδαντίδης ἐγραμμάτευ-
[εν· Ἀνθεστηριῶνο]ς τετράδι μετ' εἰκάδας, ὀγδόηι τῆς πρυ-
[τανείας· βουλὴ ἐμ βο]υλευτηρίωι

ANTHESTERION

The prytany number establishes the restoration to give Anthesterion.

Restoration to give Anthesterion 27:
A meeting of the Ekklesia in 302/1 B.C., *IG* II² 501, lines 5–9. Meritt (*Hesp* 1935, p. 546) restores these lines to give Anthesterion 29.

Anthesterion 28

A meeting in 302/1 B.C. establishes this day as a meeting day for the Ekklesia.

IG II² 500, lines 2–6:

Ἐπὶ Νικοκλέους ἄρχοντος ἐπὶ τῆ[ς Ο]-
ἰνεῖδος ὀγδόης πρυ[ταν]είας ἧι Ν[ίκ]-
ων Θεοδώρου Πλωθεὺ[ς ἐ]γραμμάτευ[ε]-
ν· τρίτηι μετ' εἰκάδας, ἑβδόμει καὶ ε-
ἰκοστῆι τῆς πρυτανεία[ς]

The prytany number establishes that the month is Anthesterion.

Restoration to give Anthesterion 28:
A meeting of the Ekklesia in 144/3 B.C., *Inscriptions de Délos* 1507, lines 37–39.

Anthesterion 29

No indisputable evidence as to the nature of this day survives.

Restorations to give Anthesterion 29:
A meeting of the Ekklesia in 327/6 B.C., *IG* II² 356, lines 1–8 as restored by Meritt, *Year*, pp. 98-99. Kirchner restored these lines to give Maimakterion 29.
A meeting of the Ekklesia in 302/1 B.C., *IG* II² 501, lines 5–9 as restored by Meritt, *Hesp* 1935, p. 546. Also *IG* II² 562, lines 1–4 as restored by Schweigert, *Hesp* 1940, p. 342. In *IG* II² 501 these lines are restored to give Anthesterion 27.

Anthesterion 30

Two meetings establish this day as a meeting day for the Ekklesia.

III. THE ATHENIAN CALENDAR

IG II² 661, lines 1–4: 267/6 B.C.

['Επ]ὶ Μενεκλέους ἄρχοντος ἐπὶ τῆς Πανδιονίδ-
[ος] ὁ[γ]δόης πρυτανείας ἧι Θεόδωρος Λυσιθέου
[Τρ]ι[κ]ο[ρ]ύσιος ἐγραμμάτευεν· Ἀνθεστηριῶνος
[ἕνει κ]αὶ νέαι· ἐκκλησία

IG II² 832, lines 1–5: 229/8 B.C.

Ἐπὶ Ἡλιοδώρου ἄρχοντος ἐπὶ τῆς Πανδιονίδος ὀγδόης [πρ]υτανείας ἧι Χαρίας Καλλίου Ἀθμονε[ὺ]ς ἐγραμ-
[μάτε]υεν· δήμου ψ[η]φίσματα· Ἀνθεστηριῶνο[ς ἕν]ει καὶ
νέαι, ἐνάτει καὶ εἰκοστῆι τῆς πρυταν[ε]ί[ας· ἐ]κκλη-
σία

Restoration to give Anthesterion 30:

A meeting of the Ekklesia in 244/3 B.C., *IG* II² 797, lines 3–6. Meritt (*Hesp* 1935, p. 555) restored these lines to give Boedromion 30.

Anthesterion—Summary

The Mysteries at Agrai, often referred to as the Lesser Mysteries, are attested for Anthesterion: Plutarch, *Demetrios* 26.1, ἀλλὰ τὰ μικρὰ τοῦ Ἀνθεστηριῶνος ἐτελοῦντο, τὰ δὲ μεγάλα τοῦ Βοηδρομιῶνος. Mommsen (*Feste*, p. 406) suggested that the central day of the Mysteries should be Anthesterion 20. He based this suggestion on the similarity of the truce of free passage for the Lesser Mysteries to that of the Eleusinian Mysteries. The truce for the Lesser Mysteries lasted from Gamelion 16 to Elaphebolion 10.[47] He assumed that the Mysteries at Agrai occurred at about the same time, within the period of the truce, as did the Eleusinian Mysteries.

Michael Jameson has recently (*BCH* 89, 1965, pp. 159–172) suggested numerous links between the Mysteries at Agrai and the cult of Zeus Meilikhios. If he is correct in this, the Mysteries at Agrai should be associated with the Diasia, the festival of Zeus Meilikhios celebrated on Anthesterion 23. The Mysteries at Agrai probably included several days,[48] and from the calendric study of Anthesterion

[47] *IG* I² 6, lines 76–87.
[48] Mommsen, *Feste*, p. 406.

ANTHESTERION

supra there is a period of seven days (Anthesterion 20–26) for which no meetings are attested. Unfortunately it is not possible to determine more precisely on which of these days the Mysteries at Agrai occurred.

ELAPHEBOLION

Elaphebolion 1

This day was a monthly festival day—the Noumenia.

Elaphebolion 2

This day was a monthly festival day devoted to the Agathos Daimon.

Elaphebolion 3

This day was a monthly festival day devoted to Athena.

Elaphebolion 4

This day was a monthly festival day devoted to Herakles, Hermes, Aphrodite, and Eros.

Restorations to give Elaphebolion 4:

A meeting of the Ekklesia in 279/8 B.C., *Hesp* 1948, pp. 1-2, no. 1 (Meritt), lines 1-4. Meritt later (*Hesp* 1969, p. 110) restored these lines to give Elaphebolion 12.

A meeting of the Ekklesia in 160/59 B.C., *Inscriptions de Délos* 1497 *bis*, lines 2-5.

Elaphebolion 5

A meeting in 284/3 B.C. establishes this day as a meeting day for the Ekklesia.

IG II² 656, lines 2-3:

Ἐπὶ Ἰσαίου, Ἐλαφηβολιῶνος πέμπτει ἰσταμένου

Restoration to give Elaphebolion 5:

A meeting of the Boule in 186/5 (?) B.C., Dow, *Prytaneis*, pp. 109-110, no. 53, lines 12-15 as restored by Pritchett and Neugebauer, *Calendars*, p. 75.

ELAPHEBOLION

Elaphebolion 6

This day was a monthly festival day devoted to Artemis. See also Elaphebolion 6 in Appendix II.

Elaphebolion 7

This day was a monthly festival day devoted to Apollo.

Elaphebolion 8

This day in 346 B.C. was both a festival day and a meeting day: a festival day devoted to Asklepios and the Proagon for the City Dionysia, and a meeting day at the instigation of Demosthenes.

Aeschines 3.66-67:

Δημοσθένης... γράφει ψήφισμα... ἐκκλησίαν ποιεῖν τοὺς πρυτάνεις τῇ ὀγδόῃ ἱσταμένου τοῦ Ἐλαφηβολιῶνος μηνός, ὅτ᾽ ἦν τῷ Ἀσκληπιῷ ἡ θυσία καὶ ὁ προαγών, ἐν τῇ ἱερᾷ ἡμέρᾳ, ὃ πρότερον οὐδεὶς μέμνηται γεγονός, τίνα πρόφασιν ποιησάμενος; " Ἵνα," φησίν, "ἐὰν παρῶσιν ἤδη οἱ Φιλίππου πρέσβεις, βουλεύσηται ὁ δῆμος ὡς τάχιστα περὶ τῶν πρὸς Φίλιππον."

This passage in Aeschines is revealing as to the nature of the relationship of meeting and festival days. Demosthenes felt that the presence of Philip's ambassadors required immediate action, and therefore he arranged a meeting of the Ekklesia for the day, despite the fact that it was a festival day. Aeschines criticizes Demosthenes for doing this, a thing ὃ πρότερον οὐδεὶς μέμνηται γεγονός. It is clear from this passage that not holding meetings on festival days was a matter of convention and tradition, not of religious or civil law. Had it been a matter of religious or civil law, Aeschines would surely have accused Demosthenes of ἀσέβεια or παρανομία. For a similar situation see Hekatombaion 12.

A meeting of the Iobakkhoi occurred on this day in the late second century A.D.

IG II^2 1368, lines 2-3:

Ἐπὶ ἄρχοντος Ἀρ(ρίου) Ἐπαφροδείτου, μηνὸς Ἐλαφηβολιῶνος ἧ ἐσταμένου, ἀγοράν

III. THE ATHENIAN CALENDAR

A meeting of this religious association naturally occurred shortly before the great festival of Dionysos, the City Dionysia.

Restoration to give Elaphebolion 8:

A meeting of the Ekklesia in 326/5 B.C., *IG* II² 359, lines 2–7. In view of the passage of Aeschines cited *supra*, the widespread acceptance of this restoration (Dinsmoor, *Archons*, p. 372, Pritchett and Neugebauer, *Calendars*, p. 54, and Meritt, *Year*, pp. 101–102) is rather surprising.

Elaphebolion 9

Four meetings establish this day as a meeting day for the Ekklesia.

IG II² 647, lines 1–7:   295/4 B.C.

['Επὶ Νι]κοστράτου ἄρχοντος [ἐ]-
[πὶ τῆς] Δημητ[ρ]ιάδος ἐνάτη[ς π]-
[ρυταν]είας ἧ[ι] Δωρόθεος Ἀρ[ισ]-
[τομάχ]ου Φαληρεὺς ἔγρα[μμάτ]-
[ευεν· Ἐ]λαφη[βο]λιῶνος ἐ[νάτηι]
[ἱσταμ]ένο[υ, πέμπ]τει κ[αὶ δεκ]-
[άτει τῆς πρυτα]ν[ε]ί[ας]

Hesp 1948, pp. 3–4, no. 3 (Meritt),   244/3 B.C.
lines 2–5:

Ἐπὶ Κυδήνορος ἄρχοντος ἐπὶ τῆς Ἐρεχθεῖδος ἐνάτης πρυτανείας ἧι Πολυκτήμων Εὐκτιμένου Εὐπυρίδης ἐγραμ-
μάτευεν· Ἐλαφηβολιῶνος ἐνάτηι ἱσταμένου, ἑβδόμηι καὶ
δεκάτηι τῆς πρυτανείας· ἐκκλ[η]σία κ[υ]ρία

Hesp 1934, pp. 14–18, no. 17   171/0 B.C.
(Meritt) (See also *Hesp* 1946, pp. 198–201),
lines 43–46:

['Ε]πὶ Ἀντιγένου ἄρχοντος ἐπ[ὶ τῆς _ca. 8_ ἐνάτης]
[πρυτα]-

124

ELAPHEBOLION

[ν]είας ἔι Σ[ώσανδρος Σωσικράτους] 'Αλω[πεκῆθεν]
[ἐγραμ]μά-
τευεν·['Ελα]φηβολιῶνος ἐ[ν]ά[τε]ι ἱσταμένου,[ὀγ]δό[ει καὶ]
[δεκά]-
[τει] τῆς πρυτα[νείας· ἐκ]κλησία ἐν τῶι θ[εάτρ]ωι

IG II² 1008, lines 49-50: 118/7 B.C.

'Επὶ Ληναίου ἄρχοντος ἐπὶ τῆς Αἰγεῖδος ἐνάτ[η]ς
πρυτανείας ᾗ 'Ισίδ[ωρο]ς 'Απολ[λ]ωνίου Σκαμβωνίδης
ἐγραμμά[τευ]-
εν· 'Ελαφηβολιῶνος ἐνάτῃ ἱσταμένου, [ἐ]νάτῃ (τῆ)ς
πρυταν[ε]ίας· [ἐκκ]λησία ἐν τῶι θεάτρωι

Mommsen (*Feste*, pp. 428-437) assigned the first day of the City Dionysia to this day. No ancient source gives a specific day as the first day of the Dionysia, the day which featured the procession escorting Dionysos to the theater. Mommsen's dating of Elaphebolion 9 as the first day of the City Dionysia was accepted by Deubner (*Feste*, p. 142) and by Pickard-Cambridge (*The Dramatic Festivals of Athens*[1], Oxford, 1953, p. 64).

It should be emphasized that no ancient source confirms Mommsen's dating. In fact, one ancient source directly contradicts it. The scholiast to Aeschines 3.67 indicates that the City Dionysia came a "few" days after the Proagon.

ἐγίγνοντο πρὸ τῶν μεγάλων Διονυσίων ἡμέραις ὀλίγαις
ἔμπροσθεν ἐν τῷ ᾠδείῳ καλουμένῳ τῶν τραγῳδῶν ἀγὼν καὶ
ἐπίδειξις ὧν μέλλουσι δραμάτων ἀγωνίζεσθαι ἐν τῷ θεάτρῳ·
δι' ὅ ἑτοίμως προαγὼν καλεῖται.

The Proagon occurred on Elaphebolion 8, and from the scholion we would expect the City Dionysia to begin, at the earliest, on Elaphebolion 10 (three days later, counting inclusively). Ferguson (*Hesp* 1948, pp. 134-135) has accepted this scholion as valid, and has argued persuasively that the City Dionysia began on Elaphebolion 10. His conclusion seems well justified,[49] especially in that it does not contradict the slight evidence which we have.

[49] This conclusion has been accepted by Gould and Lewis in their revision of Pickard-Cambridge's *The Dramatic Festivals of Athens*, p. 65. See also *Hesp* 1954, pp. 307-308.

III. THE ATHENIAN CALENDAR

Restorations to give Elaphebolion 9:

A meeting of the Ekklesia in 307/6 B.C., *IG* II² 460, lines 1-5. Also *IG* II² 461, lines 1-6. Also *IG* II² 462, lines 1-6 as restored by Meritt, *Hesp* 1963, p. 437. *IG* II² 460 is restored by Pritchett and Meritt (*Chronology*, pp. 17-18) to give Thargelion 2, and by Meritt (*Year*, pp. 177-178) to give Skirophorion 3. *IG* II² 461 is restored by Pritchett and Meritt (*Chronology*, p. 17) to give Elaphebolion 20, and by Meritt (*Year*, p. 177) to give Elaphebolion 25.

A meeting of the Ekklesia in 298/7 B.C., *Hesp* 1940, pp. 80-83, no. 13 (Meritt), lines 1-6 as restored by Meritt, *Hesp* 1969, pp. 107-108. Meritt originally restored these lines to give Elaphebolion 22.

A meeting of the Ekklesia in 295/4 B.C., *IG* II² 646, lines 1-5.

A meeting of the Ekklesia in 275/4 B.C., *Hesp* 1964, pp. 170-171, no. 25 (Meritt), lines 1-4.

A meeting of the Ekklesia in 244/3 B.C., Pritchett and Meritt, *Chronology*, pp. 23-27, lines 1-5 as restored by Meritt, *Hesp* 1948, p. 4.

A meeting of the Ekklesia in 139/8 (?) B.C., *Hesp* 1960, pp. 76-77, no. 154 (Meritt), lines 1-4.

Elaphebolion 10

This was the first day of the City Dionysia, the day of the procession. See the discussion for Elaphebolion 9.

The Iobakkhoi in the second century A.D. made a sacrifice and libation to Dionysos on this day.

IG II² 1368, lines 117-121:

ὁ δὲ ἀρχί-
βαχκος θυέτω τὴν θυσίαν τῷ
θεῷ καὶ τὴν σπονδὴν τιθέτω
κατὰ δεκάτην τοῦ Ἐλαφηβολιῶνος μηνός

Such a sacrifice by this religious association naturally would occur on the first day of the City Dionysia.

The sacrificial calendar of the Marathonian Tetrapolis prescribes an offering to Ge (?) for this day.

ELAPHEBOLION

IG II² 1358, Column II, lines 17-18:
Ἐλα[φη]βολιῶνος δεκάτηι ἱσταμένο· [Γῆι ἐπὶ τῶι]
μαν[τε]ίωι τράγος παμμέλας Δ Γ. ἱε[ρώσυνα ⊢.]

Restoration to give Elaphebolion 10:
A meeting of the Ekklesia in 189/8 B.C., Dow, *Prytaneis*,
pp. 91-92, no. 41, lines 25-26 as re-edited with a new fragment by
Meritt, *Hesp* 1957, pp. 63-64, no. 17.

Elaphebolion 11

If we accept Ferguson's dating (see Elaphebolion 9), this would
be the second day of the City Dionysia, presumably devoted to the
dithyrambic or dramatic competitions. The restoration of the whole
program of these competitions is very hypothetical.

Restoration to give Elaphebolion 11:
A meeting of the Ekklesia in 280/79 B.C., *IG* II² 670, lines 1-4.
Kirchner later (*IG* II² 670 addenda) restored these lines to give
Skirophorion 12. These lines are restored by Meritt (*Hesp* 1969,
pp. 109-110) to give Gamelion 16. Meritt earlier (*Hesp* 1938,
p. 106) had restored these lines to give Skirophorion 11.

Elaphebolion 12

If we accept Ferguson's dating (see Elaphebolion 9), this would
be the third day of the City Dionysia, presumably devoted to the
dramatic competitions.
A meeting of the Ekklesia occurred on this day in 319/8 B.C.

Hesp 1938, pp. 476-479, no. 31
(M. Crosby) (See also Meritt, *Year*, p. 122),
lines 1-7:

[Ἐπὶ Ἀπο]λλοδ[ώρ]ου ἄρχοντος ἐπ[ὶ τ]-
[ῆς Ἀντι]οχί[δος ἑ]βδόμης πρυτα[νε]-
[ίας καὶ ἀναγραφέ]ως Εὐκάδμου Ἀ[ν]-
[ακαι]έως· Ἐλαφ[ηβο]λιῶνος δωδεκ[ά]-
τει, τετάρτει [καὶ τ]ριακοστεῖ τ[ῆ]-
ς πρυτανείας· ἐ[κκ]λ[η]σία κατὰ ψ[ήφ]-
ισμα βουλῆς

III. THE ATHENIAN CALENDAR

This is one of the three meetings which are attested to have occurred in the course of the City Dionysia (see Elaphebolion 13 and 14). The dramatic competitions evidently could be postponed. A meeting such as this one in 319/8 B.C. may have required a postponement of only a few hours.

Restoration to give Elaphebolion 12:

A meeting of the Ekklesia in 279/8 B.C., *Hesp* 1948, pp. 1-2, no. 1 (Meritt), lines 1-4 as restored by Meritt, *Hesp* 1969, p. 110. Meritt originally restored these lines to give Elaphebolion 4.

Elaphebolion 13

If we accept Ferguson's dating (see Elaphebolion 9), this would be the fourth day of the City Dionysia, presumably devoted to the dramatic competitions.

A meeting of the Ekklesia evidently occurred on this day in 196/5 B.C.

Hesp 1936, pp. 419-428, no. 15
(Meritt), lines 1-4:

Ἐπὶ Χαρικλέους ἄρχοντος ἐπὶ τῆς Αἰγεῖδος ἐνάτης
πρυτανείας ἧι
Αἰσχρίων Εὐαινέτου Ῥαμνούσιος ἐγραμμάτευεν·
δήμου ψηφίσματα·
Ἐλαφηβολιῶνος τρίτει ἐπὶ δέκα κατὰ θεὸν δὲ ὀγδόει
καὶ εἰκοστῶι
τῆς πρυτανείας· ἐκκλησία κυρία ἐμ Πειραιεῖ

A meeting of the Ekklesia in the Piraeus would certainly have required a one-day postponement of the events of the City Dionysia.[50] There is some confusion concerning the day date of this inscription. It would appear superficially that the κατὰ θεόν date was accidentally omitted. This may, however, be only a symptom of deeper confusion. Unfortunately the evidence is insufficient to demonstrate that the confusion also affected what appears to be the κατ' ἄρχοντα date.

[50] Ferguson, *Hesp* 1948, pp. 133-135, note 46.

ELAPHEBOLION

Restoration to give Elaphebolion 13:

A meeting of the Ekklesia in 322/1 B.C., *IG* II² 372, lines 1-6 as restored by Meritt, *Year,* pp. 110-111. Schweigert (*Hesp* 1939, pp. 173-175) follows Kirchner in restoring these lines to give Elaphebolion 19.

Elaphebolion 14

According to Thucydides 4.118.11-13 the Athenians on this day in 423 B.C. voted the preliminary measures which led to the Peace of Nikias: ἔδοξεν τῷ δήμῳ... τὴν (δ') ἐκεχειρίαν εἶναι ἐνιαυτόν, ἄρχειν δὲ τήνδε τὴν ἡμέραν, τετράδα ἐπὶ δέκα τοῦ Ἐλαφηβολιῶνος μηνός. This in all probability indicates that a meeting of the Ekklesia occurred on this day.[51] Two other meetings are attested for the days of the City Dionysia (see Elaphebolion 12 and 13), and therefore this meeting does not provide a *terminus ante quem* for the festival.[52]

Elaphebolion 15

An offering for Kronos is specified for this day in a private sacrificial calendar of the first or second century A.D.

IG II² 1367, lines 23-26:

['Ε]λαφηβολιῶνος εἶ Κρόνῳ πόπανον
δωδεκόμφαλον καθήμενον ἐπι[..]
[..]σεις βοῦν χοινικιαῖον ἀνυπε[ρθέ]-
τως

No other evidence as to the nature of this day survives.

Elaphebolion 16

The sacrificial calendar of the deme Erkhia prescribes offerings for Dionysos and Semele on this day.

---

[51] Pickard-Cambridge, *The Dramatic Festivals of Athens*¹, p. 57, note 7.
[52] Ferguson, *Hesp* 1948, pp. 133-135, note 46 as against Deubner, *Feste,* p. 142.

III. THE ATHENIAN CALENDAR

Column A, lines 44-51
['E]λαφηβολιῶνο-
ς ἕκτηι ἐπὶ δέ-
κα, Σεμέληι, ἐπ-
ὶ τοῦ αὐτοῦ βω-
μοῦ, αἴξ, γυναι-
ξὶ παραδόσιμ-
ος, ἱερέας τὸ δ-
έρμα, οὐ φορά, Δ

Column Δ, lines 33-40
Ἐλαφηβολιῶν-
ος ἕκτηι ἐπὶ
δέκα, Διονύσ-
ωι, Ἐρχιᾶ, αἴξ,
παραδό: γυνα-
(α)ιξί, οὐ φορά,
ἱερέαι τὸ δέ-
ρμα, Δ ⊢⊢

These sacrifices suggest that the festival of Dionysos at Athens was still being celebrated, or had just finished.

Elaphebolion 17

No evidence as to the nature of this day survives, but refer also to Elaphebolion—Summary.

Elaphebolion 18

Aeschines attests meetings of the Ekklesia on this day and the following day in 346 B.C.

Aeschines 2.61:

Παρανάγνωθι δή μοι καὶ τὸ Δημοσθένους ψήφισμα, ἐν ᾧ κελεύει τοὺς πρυτάνεις μετὰ τὰ Διονύσια τὰ ἐν ἄστει καὶ τὴν ἐν Διονύσου ἐκκλησίαν προγράψαι δύο ἐκκλησίας, τὴν μὲν τῇ ὀγδόῃ ἐπὶ δέκα, τὴν δὲ τῇ ἐνάτῃ

Aeschines 3.68:

ἐνταῦθ' ἕτερον νικᾷ ψήφισμα Δημοσθένης... βουλεύσασθαι ... εὐθὺς μετὰ τὰ Διονύσια τὰ ἐν ἄστει, τῇ ὀγδόῃ καὶ ἐνάτῃ ἐπὶ δέκα

Elaphebolion 19

Three meetings establish this day as a meeting day for the Ekklesia.

A meeting of the Ekklesia in 346 B.C. as attested by Aeschines 2.61 and 3.68, cited *supra* for Elaphebolion 18.

ELAPHEBOLION

IG II² 345, lines 2-7:   332/1 B.C.

Ἐπὶ Νικήτου ἄρ[χοντος ἐ]πὶ τῆς Ἀντιοχίδος ὀγδόης π[ρυτανεία]ς ἧι Ἀριστόνο-
υς Ἀριστόνο[υ Ἀναγυράσι]ος ἐγραμ[μά]τ-
ευεν· Ἐλαφη[βολιῶνος ἐν]άτηι ἐπὶ δέ[κ]α,
ἑβδόμ[ηι τῆς πρυτανείας·] ἐκκλησία[ἐ]ν
[Διονύσου]

This restoration is determined by the spacing, but is confirmed by *IG* II² 346 and 347.

Hesp 1957, pp. 72-77, no. 22 (Meritt),   164/3 B.C.
lines 1-4:

Ἐπὶ Εὐεργέτου ἄρχοντος ἐπὶ τῆς Ἱπποθωντίδος
   ἐνάτης πρυτ[α]-
νείας ἧι Διονυσόδωρος Φιλίππου Κεφαλῆθεν
   ἐγραμμάτευε[ν·]
Ἐλαφηβολιῶνος ἐνάτει ἐπὶ δέκα, κατὰ θεὸν δὲ
   δεκάτει ὑστέ[ραι,]
δευτέραι καὶ εἰκοστεῖ τῆς πρυτανείας· ἐκκλησία
   ἐμ Πειρ[αιεῖ]

Restorations to give Elaphebolion 19:

A meeting of the Ekklesia in 332/1 B.C., *Hesp* 1939, pp. 26-27, no. 6 (Schweigert), lines 1-6.
A meeting of the Ekklesia in 331/0 B.C., *IG* II² 348, lines 1-5. See also Meritt, *Year*, pp. 90-91.
A meeting of the Ekklesia in 322/1 B.C., *IG* II² 372, lines 1-6. Schweigert (*Hesp* 1939, pp. 173-175, no. 4) accepts this restoration to give Elaphebolion 19, but Meritt (*Year*, pp. 110-111) restores these lines to give Elaphebolion 13.
A meeting of the Ekklesia in 280/79 B.C., *Hesp* 1941, pp. 338-339 (Schweigert), lines 1-4.

Elaphebolion 20

No indisputable evidence as to the nature of this day survives.

Restorations to give Elaphebolion 20:

A meeting of the Ekklesia in 307/6 B.C., *IG* II² 461, lines 1-6 as

III. THE ATHENIAN CALENDAR

restored by Pritchett and Meritt, *Chronology*, p. 17. These lines are restored by Kirchner to give Elaphebolion 9 and by Meritt (*Year*, p. 177) to give Elaphebolion 25. A meeting of the Ekklesia in the third century B.C., *IG* II² 857, lines 1-3.

Elaphebolion 21

Four meetings establish this day as a meeting day of the Ekklesia.

IG II² 780, lines 2-4: 252/1 B.C.

> Ἐπὶ Καλλιμήδου ἄρχοντος ἐπὶ τῆς Αἰαντίδος ἐνάτης
> πρυτανεία[ς ἧι Καλ]-
> λίας Καλλιάδου Πλωθεὺς ἐγραμμάτευεν· Ἐλαφηβολιῶνος
> δεκάτηι [ὑστέρα]-
> ι, ἐνάτηι καὶ εἰκοστῆι τῆς πρυτανείας· ἐκκλησία ἐν
> Διονύσου

IG II² 781, lines 2-5: 250/49 B.C.

> [Ἐπὶ Θερσιλόχου ἄρχοντος ἐπὶ τ]ῆς[....¹¹......]ἐνάτης π-
> [ρυτανείας ἧι Διόδοτος Διογν]ήτου Φρεάρριος ἐγραμμά-
> [τευεν· Ἐλαφηβολιῶνος δεκάτηι] ὑστέραι, τετάρ[τ]ηι καὶ εἰ-
> [κοστῆι τῆς πρυτανείας· ἐκκλησ]ία ἐν Διονύσου

The restoration to give Elaphebolion is established by the number and day of the prytany.

IG II² 896, lines 2-5: 186/5 B.C.

> [Ἐπὶ Ζ]ωπύρου ἄρχοντος ἐπὶ τῆς Πτολε[μα]ιίδος δεκ[άτ]ης
> [πρυ]-
> [τανε]ίας ἧι Μεγάριστος Πύρρου Αἰξωνε[ὺ]ς ἐγραμ-
> μ[ά]τευεν·
> [Ἐλαφ]ηβολιῶνος δεκάτει ὑστέραι, τετάρτει τῆς
> πρυτανεί-
> [ας· ἐκκλ]ησία ἐν Διονύσου

lines 29-32:

> Ἐπὶ Ζωπύρου ἄρχοντος ἐπὶ τῆς Πτολεμαιίδος δεκάτης
> πρυτανεί-

ELAPHEBOLION

ας ἧι Μεγάριστος Πύρρου Αἰξωνεὺς ἐγραμμάτευεν·
Ἐλαφηβολιῶ-
νος δεκάτει ὑστέραι, τετάρτει τῆς πρυτανείας·
ἐκκλησία ἐν Διο-
νύσου

Hesp 1935, pp. 71–81, no. 37 (Dow), as   128/7 B.C.
re-edited with new fragments by
Meritt, Hesp 1946, pp. 201–213, no. 41,
lines 103–104:

κατὰ τὸ ψήφισμα ὃ Τίμαρχος Ἐπηρατίδου Σφ[ήττιος εἶπεν]
[.....$\overset{ca.\ 12}{.....}$.... Ἐλαφη]βο[λι]ῶνος δ[εκάτ]ει ὑστέραι,
μ[ιᾶι]
καὶ εἰκοστῆι τῆς πρυτανείας· ἐκκλησία ⟦ι⟧ ἐν [τῶι θεάτρωι]

Restorations to give Elaphebolion 21:
A meeting of the Ekklesia in 335/4 B.C., *IG* II^2 332, lines 1–5.
A meeting of the Ekklesia in 254/3 B.C., *IG* II^2 697, lines 1–5.
These lines are restored by Dow (*Hesp* 1963, pp. 352–356) to give
Skirophorion 21. See also *Hesp* 1969, pp. 433–434.
A meeting of the Ekklesia in 250/49 B.C., *IG* II^2 780, lines 26–28.
A meeting of the Ekklesia in the second century B.C., *IG* II^2 991,
line 1.

Elaphebolion 22

Two meetings establish this day as a meeting day for the Ekklesia.

IG II^2 680, lines 1–4:   249/8 B.C.

[Ἐ]πὶ Πολυεύκτου ἄρχοντος ἐπὶ τῆς Αἰγεῖδος ἐνάτης πρ-
[υ]τανείας ἧι Χαιρεφῶν Ἀρχεστράτου Κεφαλῆθεν ἐγρα-
[μ]μάτευεν· Ἐλαφηβολιῶνος ἐνάτει μετ' εἰκάδας, τριακο-
[σ]τεῖ τῆς πρυτανείας· ἔδοξεν τῶι δήμωι

IG II^2 967, lines 1–5:   145/4 B.C.

[Ἐπ]ὶ Μητροφάνου ἄρχοντος ἐπὶ τῆς Ἀκαμαντίδος
δεκάτης πρυτα-
νείας ἧι Ἐπιγένης Μοσχίωνος Λαμπτρεὺς ἐγραμμάτευεν·
ἀντι-

III. THE ATHENIAN CALENDAR

γραφεὺς Δημοκράτης Δημοκράτου Κυδαθηναιεύς·
'Ελαφηβολιῶνο[ς]
ἐνάτει μετ' εἰκάδας κατ' ἄρχοντα, κατὰ θεὸν [δ]ὲ
[Μ]ουνιχιῶνος δωδε[κά]-
τει, δωδεκάτει τῆς πρυτανείας· ἐκκλησία κυρία ἐν
τῶι θεάτρωι

Restorations to give Elaphebolion 22:
A meeting of the Ekklesia in 298/7 B.C., *Hesp* 1940, pp. 80-83, no. 13 (Meritt), lines 1-6. Meritt later (*Hesp* 1969, pp. 107-108) restored these lines to give Elaphebolion 9.
A meeting of the Ekklesia in 239/8 B.C., *Hesp* 1938, pp. 123-126, no. 25 (Meritt), lines 1-5.

Elaphebolion 23

A meeting in 190/89 B.C. establishes this day as a meeting day for the Ekklesia.

Pritchett and Meritt, *Chronology*,
pp. 121-123, lines 1-4:

Ἐπὶ Δημητρίου ἄρχοντος ἐπὶ τ[ῆς Ἀντιοχίδος ἐνάτης]
[πρυτα]-
νείας· Ἐλαφηβολιῶνος ὀγδόει μετ' ε̣[ἰκάδας, κατὰ]
[θεὸν δὲ]
τετράδι μετ' εἰκάδας, ὀγδόει καὶ εἰκοσ[τὲι τῆς]
[πρυτανεί]-
ας· ἐκκλησία ἐμ Πειραιεῖ

Elaphebolion 24

No evidence as to the nature of this day survives.

Elaphebolion 25

Aeschines states that a meeting of the Ekklesia occurred on this day in 346 B.C.

ELAPHEBOLION

Aeschines 3.73:

εἰς δὲ τὴν ἐκκλησίαν τὴν τῇ ἕκτῃ ['Ελαφηβολιῶνος] προκαθεζόμενος βουλευτὴς ὢν ἐκ παρασκευῆς, [Δημοσθένης] ἔκδοτον Κερσοβλέπτην μετὰ Φιλοκράτους ἐποίησε.

Aeschines 2.90:

Δημοσθένης δ' ἐν τῷ δήμῳ προήδρευε τούτου τοῦ μηνός ['Ελαφηβολιῶνος], εἷς ὢν τῶν πρέσβεων, ἕκτῃ φθίνοντος.

Refer also to Elaphebolion 25 in Appendix II.

Restoration to give Elaphebolion 25:

A meeting of the Ekklesia in 307/6 B.C., *IG* II² 461, lines 1–6 as restored by Meritt, *Year*, p. 177. These lines are restored by Kirchner to give Elaphebolion 9, and by Pritchett and Meritt (*Chronology*, p. 17) to give Elaphebolion 20.

Elaphebolion 26

No indisputable evidence as to the nature of this day survives.

Restoration to give Elaphebolion 26:

A meeting of the Ekklesia in 272/1 B.C., *IG* II² 704, lines 1–7 as restored by Meritt, *Hesp* 1957, pp. 56–57. These lines are restored by Kirchner to give Elaphebolion 27.

Elaphebolion 27

No indisputable evidence as to the nature of this day survives.

Restoration to give Elaphebolion 27:

A meeting of the Ekklesia in 272/1 B.C., *IG* II² 704, lines 1–7. These lines are restored by Meritt (*Hesp* 1957, pp. 56–57) to give Elaphebolion 26.

Elaphebolion 28

No evidence as to the nature of this day survives.

III. THE ATHENIAN CALENDAR

Elaphebolion 29

A meeting in 197/6 B.C. establishes this day as a meeting day for the Ekklesia.

IG II² 850, lines 1-3:

Ἐπὶ Διονυσίου ἄρχ[οντ]ος· Ἐλαφηβολιῶνος δευτέραι μετ' εἰκάδας· ἐκκλησί[α]
ἐμ Πειραιεῖ

Restoration to give Elaphebolion 29:
A meeting of the Ekklesia in 305/4 (?) B.C., *IG* II² 703, lines 2-6.

Elaphebolion 30

Three meetings establish this day as a meeting day for the Ekklesia.

IG II² 354, lines 2-7: 328/7 B.C.

[Ἐπ' Εὐθυκρίτου ἄρ]χοντος, ἱερείως δὲ Ἀνδρο-
[κλέους ἐκ Κεραμ]έων· ἐπὶ τῆς Ἀντιοχίδος ὀγ-
[δόης πρυτανεία]ς ἧι Πυθόδηλος Πυθοδήλου
[Ἁγνούσιος ἐγρα]μμάτευεν· ἔνηι καὶ νέαι π-
[ροτέραι, εἰκοστ]ῆι τῆς πρυτανείας· ἐκκλησ-
[ία]

The number of the prytany establishes that the month is Elaphebolion in an ordinary year.

IG II² 336b, lines 5-7: 320/19 B.C.

[Ἐλαφ]ηβολιῶνος ἔνε[ι καὶ νέαι, ἕκτει κα]-
[ὶ εἰκ]οστεῖ τῆς πρυτ[ανείας ἧι ...⁷...]
[..⁵..]νεὺς ἐγραμμά[τευεν· ἐκκλησία]

IG II² 662 as re-edited with a new 286/5 B.C.
fragment by M. Lethen, *Hesp* 1957,
pp. 29-30, no. 2, lines 1-3:

Ἐπὶ Διοκλέους ἄρχον[τος ἐπὶ τῆς Κεκ]ροπίδος
ἐνάτης π-
ρυτανείας· Ἐλαφηβολ[ιῶνος ἕνει καὶ] νέαι, τριακοστεῖ
τῆς πρυτανε[ίας· ἐ]κκλ[ησία κυρία]

136

ELAPHEBOLION

Restoration to give Elaphebolion 30:

A meeting of the Ekklesia in 319/8 B.C., *IG* II² 388, lines 1–6. Crosby (*Hesp* 1938, pp. 478–479) restores these lines to give Mounichion 12.

Elaphebolion—Summary

The relatively large number of meetings attested for this month is remarkable. The large number of meetings may partially be explained by the fact that Elaphebolion fell in early spring.

The extent of the City Dionysia and the Pandia which followed[53] has been a subject of constant discussion since the studies by Dutoit[54] and Mommsen[55] in 1898. Ferguson (*Hesp* 1948, pp. 134–135) contributed two valuable points to the discussion: first, the first day of the festival was probably Elaphebolion 10; secondly, meetings of the Ekklesia did occur during the course of the City Dionysia (see Elaphebolion 12 and 13), and therefore the meeting of the Ekklesia on Elaphebolion 14 does not provide a *terminus ante quem* for the festival. In view of this, new consideration should be given to the testimony of Aeschines (3.68, cited *supra* for Elaphebolion 18), who states that the meeting of the Ekklesia on Elaphebolion 18 occurred "immediately" after the Dionysia. From this evidence and from the calendric study of Elaphebolion *supra* it would appear that the City Dionysia included the days Elaphebolion 10–16, and that the Pandia occurred on Elaphebolion 17.

[53] Photios Πάνδια, and Deubner, *Feste*, p. 176.
[54] Dutoit, *Zur Festordnung der grosser Dionysien*.
[55] Mommsen, *Feste*, pp. 428–448.

MOUNICHION

Mounichion 1

This day was a monthly festival day—the Noumenia.

Mounichion 2

This day was a monthly festival day devoted to the Agathos Daimon. Legal proceedings are attested for this day in 325/4 B.C.

IG II² 1629, lines 206-212:

[τοὺ]ς θεσμοθέτας παρα-
[πλ]ηρῶσαι δικαστήρια εἰς
[ἕν]α καὶ διακοσίους τῶι
[στ]ρατηγῶι τῶι ἐπὶ τὰς συμ-
[μ]ορίας ἡιρημένωι ἐν τῶι
[Μ]ουνιχιῶνι μηνὶ τῆι δευ-
[τ]έραι ἱσταμένου

Mounichion 3

This day was a monthly festival day devoted to Athena. Aeschines, 2.91-92, attests a meeting of the Boule on this day in 346 B.C.: καί μοι λέγε τὸ τῆς βουλῆς ψήφισμα. [ΨΗΦΙΣΜΑ]... Ἀκούετε ὅτι Μουνιχιῶνος ἐψηφίσθη τρίτῃ ἱσταμένου. For a financial transaction on this day, see Mounichion 3 in Appendix I.

Mounichion 4

This day was a monthly festival day devoted to Herakles, Hermes, Aphrodite, and Eros. Two of these deities are involved in the religious activities attested for this day.

A "festival" or "banquet" for Eros is designated for this day on an inscription dated to the middle of the fifth century B.C.

Hesp 1932, pp. 43-44 (Broneer), lines 1-3:

MOUNICHION

τôι ῞Εροτι hε έορτὲ
[τ]ετράδι hισταμέν[ο]
Μονιχιôν[ο]ς μεν[ός]

There is no evidence that this inscription concerns a state festival. By itself it is not sufficient to indicate that the day was a state festival day. This inscription may be the record of a private religious association. The sacrificial calendar of the deme Erkhia designates an offering to the Herakleidai on this day.

Column B, lines 40–44

Μουνιχιῶνο-
ς τετράδι ἱσ-
ταμένου, Ἡρα-
κλείδαις, οἶ-
ς, Ἐρχιᾶ, Δ ⊢⊢

For a financial transaction on this day, see Mounichion 4 in Appendix I.

Restoration to give Mounichion 4:
A meeting of the Ekklesia in 128/7 B.C., *Hesp* 1935, pp. 71–81, no. 37 (Dow), lines 115–116 as restored by Meritt, *Hesp* 1946, pp. 201–213, no. 41.

Mounichion 5

Legal proceedings are attested for this day in 325/4 B.C.

IG II² 1629, lines 206–213:

[τοὺ]ς θεσμοθέτας παρα-
[πλ]ηρῶσαι δικαστήρια εἰς
[ἕν]α καὶ διακοσίους τῶι
[στ]ρατηγῶι τῶι ἐπὶ τὰς συμ-
[μ]ορίας ἡιρημένωι ἐν τῶι
[Μ]ουνιχιῶνι μηνὶ τῆι δευ-
[τ]έραι ἱσταμένου καὶ τῆι
[π]έμπτηι ἱσταμένου

No other evidence as to the nature of this day survives.

III. THE ATHENIAN CALENDAR

Mounichion 6

Plutarch, *Thes.* 18.1, states that on this day the young girls went in procession to the Delphinion: ἕκτῃ μηνὸς... ἱσταμένου Μουνυχιῶνος, ᾗ καὶ νῦν ἔτι τὰς κόρας πέμπουσιν ἱλασομένας εἰς Δελφίνιον. Deubner (*Feste*, p. 201) associates this procession with the festival named Delphinia. E. Pfuhl (*De Atheniensium Pompis Sacris*, p. 79) suggests that this procession was in honor of Artemis rather than Apollo. He points out that the sixth day was regarded as the birthday of Artemis and that both Artemis and Apollo were worshipped in the Delphinion.

A meeting of the association of the Soteriastai occurred on this day ca. 37/6 B.C.

IG II² 1343, lines 6-7:

Ἐπὶ Θεοπίθου ἄρχοντος, Μουνιχιῶνος
ἕκτῃ

This meeting of the Soteriastai should be associated with the procession to the Delphinion. The goddess Soteira worshipped by the Soteriastai was Artemis because the inscription was found in the sanctuary of Artemis Soteira. This meeting of Artemis' devotees confirms Pfuhl's claim that it was Artemis, not Apollo, who was being worshipped on this day. This meeting of the Soteriastai on the festival day of their guardian deity does not indicate that the day was a meeting day.

For a financial transaction on this day, see Mounichion 6 in Appendix I.

Mounichion 7

This day was a monthly festival day devoted to Apollo.

Mounichion 8

This day was a monthly festival day devoted to Poseidon and Theseus.

Restoration to give Mounichion 8:

A meeting of the Ekklesia in 320/19 B.C., *Hesp* 1944, pp. 234-241, no. 6 (Meritt), lines 2-6 as restored by Meritt, *Year*, pp. 119-120.

140

MOUNICHION

Meritt originally (*Hesp* 1944, pp. 234–241) and then again later (*Hesp* 1963, pp. 425–432) restored these lines to give Gamelion 10. Dow (*HSCP* 67, 1963, pp. 67–75) restored these lines to give Gamelion 6.

Mounichion 9
No evidence as to the nature of this day survives.

Mounichion 10
Except for a financial transaction (see Mounichion 10 in Appendix I), no indisputable evidence concerning the nature of this day survives.

Restorations to give Mounichion 10:
A meeting of the Ekklesia in 302/1 B.C., *Hesp* 1935, pp. 37–38, no. 6 (Oliver), lines 1–6.
A meeting of the Ekklesia in 175/4 B.C., *IG* II² 905, lines 1–5.
Pritchett and Meritt (*Chronology*, p. 121) restore these lines to give Mounichion 11.

Mounichion 11
Four meetings establish this day as a meeting day for the Ekklesia.

Pritchett and Meritt, *Chronology*,   187/6 B.C.
pp. 117–118, lines 2–5:

['Επὶ Θεοξένου ἄρ]χοντος ἐπὶ τῆς 'Ερε[χθεῖδος δεκάτης]
[πρυτ]ανείας ἧι Βιοτέλης Λευκίου Π[εριθοίδης ἐγραμμά]-
[τε]υεν· Μουνυχιῶνος ἑνδεκάτηι, τ[ρίτηι καὶ δεκάτηι]
[τῆ]ς πρυτανείας· ἐκκλησία κυρία [ἐν τῶι θεάτρωι]

IG II² 897, lines 1–6:   185/4 B.C.

'Επὶ Εὐπολέμου ἄρχοντος ἐπὶ τῆς Πτολεμαιίδο[ς δε]-
[κ]άτης πρυτανείας ἧι Στρατόνικος Στρατονίκ[ου 'Α]-
[μα]ξαντεὺς ἐγραμμάτευεν· Μουνιχιῶνος ἐν[δεκά]-
τει· βουλὴ ἐμ βουλευτηρίωι σύγκλητος στρατ[ηγῶν]
παραγγειλάντων καὶ ἀπὸ βουλῆς ἐκκλησία [κυρία]
ἐν τῶι θεάτρωι

III. THE ATHENIAN CALENDAR

See also *IG* II² 898.

IG II² 996, lines 1–3: mid-second century B.C.

['Επὶ ..ᶜᵃ: ⁷... ἄρχοντος ἐ]πὶ τῆς Πτολεμ[αίδος δεκά]-
[της πρυτανείας· Μουν]ιχιῶνος ἐνδε[κάτει, ἐνδε]-
[κάτει τῆς πρυτανεία]ς· ἐκκλησία κυρ[ία]

Josephos *AJ* 14.8.5:   106/5 B.C.

Ἐπὶ Ἀγαθοκλέους ἄρχοντος Εὐκλῆς Ξενάνδρου Αἰθαλίδης
ἐγραμμάτευε, Μουνυχιῶνος ἑνδεκάτῃ τῆς πρυτανείας,
ἐκκλησίας γενομένης ἐν τῷ θεάτρῳ

Restorations to give Mounichion 11:
A meeting of the Ekklesia in 251/0 B.C., *IG* II² 768, lines 2–6.
A meeting of the Ekklesia in 188/7 B.C., *IG* II² 891, lines 1–3.
These lines were restored by Meritt (*AJP* 78, 1957, p. 381) to give Mounichion 19, and later (*Year*, p. 155) to give Metageitnion 19.
A meeting of the Ekklesia in 175/4 B.C., *IG* II² 905, lines 1–5 as restored by Pritchett and Meritt, *Chronology*, p. 121. These lines are restored by Kirchner to give Mounichion 10.
A meeting of the Ekklesia in 173/2 B.C., *Hesp* 1947, p. 163, no. 61 (Meritt), lines 1–4 as restored by Meritt, *Hesp* 1957, p. 39.

Mounichion 12

A meeting in 166/5 B.C. establishes this day as a meeting day for the Ekklesia.

IG II² 947, lines 9–12:

['Επὶ Ἀχαιοῦ] ἄρχοντος ἐπὶ τῆς[- - -]ς ἑνδεκάτης
πρυτ[α]ν[ε]ία[ς] ἧι ['Ηρ]-
[ακλέων Ναννʹ]άκου Εὐπυρίδης ἐγρ[αμμάτε]υεν· Μουνιχιῶνος [δ]ωδ[εκ]άτ[ηι,]
[κατὰ θεὸν δὲ] Θαργηλιῶνος [δωδ]ε[κάτηι], δωδεκάτη[ι τῆ]ς πρυ[τανείας·]
[ἐκκλησία κ]υρία ἐν τῶι θεάτρωι

Restorations to give Mounichion 12:
A meeting of the Ekklesia in 319/8 B.C., *IG* II² 388, lines 1–6 as

restored by Crosby, *Hesp* 1938, pp. 478–479. Kirchner restored these lines to give Elaphebolion 30.

A meeting of the Ekklesia in 252/1 B.C., *IG* II² 777, lines 1–4.

A meeting of the Ekklesia in 244/3 B.C., *Hesp* 1938, pp. 114–115, no. 21 + *IG* II² 766, lines 1–4 as restored by Meritt, *Hesp* 1948, pp. 5–7. Meritt also restored as "possible" dates Boedromion 12 (*Hesp* 1938, pp. 114–115), Posideon 12 (*Year*, p. 148), and Pyanopsion 16 (*Χαριστήριον εἰς Ἀναστάσιον Κ. Ὀρλάνδον*, Vol. I, 1965, pp. 193–197).

Mounichion 13

No evidence as to the nature of this day survives.

Mounichion 14

A meeting of the Ekklesia of the Athenian cleruchs on Delos occurred on this day in 147/6 B.C.

Inscriptions de Délos 1503, lines 1–3:

['Ἐπὶ Ἄρχ]οντος ἄρχοντος, Μουνιχιῶ[νος]
[τετρά]δι ἐπὶ δέκα, ἐκ(κ)λησία κυρία ἐν τῶι
[ἐκκ]λησιαστηρίωι

This meeting of cleruchs is not sufficient evidence to establish this day as a meeting day in Athens.

Restoration to give Mounichion 14:

A meeting of the Boule in 336/5 B.C., *IG* II² 330, lines 29–31.

Mounichion 15

No evidence as to the nature of this day survives.

Mounichion 16

This day was a festival day—the day of the Mounichia in honor of Artemis. Three ancient sources establish the date:

III. THE ATHENIAN CALENDAR

Plutarch *Mor*. 349F: τὴν δ' ἕκτην ἐπὶ δέκα τοῦ Μουνιχιῶνος Ἀρτέμιδι καθιέρωσαν, ἐν ᾗ τοῖς Ἕλλησι περὶ Σαλαμῖνα νικῶσιν ἐπέλαμψεν ἡ θεὸς πανσέληνος

Photios: ἀμφιφόων· Φιλόχορος ἐν τῇ περὶ ἡμερῶν· ἕκτῃ ἐπὶ δέκα· καὶ τοὺς καλουμένους δὲ νῦν ἀμφιφῶντας ταύτῃ τῇ ἡμέρᾳ πρῶτον ἐνόμισαν οἱ ἀρχαῖοι φέρειν εἰς τὰ ἱερὰ τῇ Ἀρτέμιδι

Suda: ἀνάστατοι· οἱ δὲ ἀμφιφῶντες γίνονται Μουνυχιῶνος μηνὸς ἕκτῃ ἐπὶ δέκα, οἳ καὶ εἰς τὸ Μουνυχίας ἱερὸν τῆς Ἀρτέμιδος κομίζονται

A meeting of the Ekklesia is attested for this day in 296/5 B.C.

IG II² 644, lines 1–6:

Ἐπὶ Νικίου ἄρχοντος ὑστέρ[ου ἐπὶ]
τῆς Ἀκαμ[α]ντίδος τετάρτης π[ρυτα]-
[νε]ίας ἧι Ἀ[ν]τι[κρ]άτης Κρατίν[ου Ἀζ]-
[ην]ι[εὺς ἐγραμμ]άτευε· Μουνιχ[ιῶ]-
[ν]ος ἕκ[τηι ἐπὶ δέ]κ[α], ἑβδόμη[ι τῆς]
[π]ρυτα[νείας· ἐκκ]λη[σ]ία

This is an instance of a meeting of the Ekklesia occurring on a festival day.

Mounichion 17

A meeting of the association of the Thiasotai occurred on this day in 278/7 B.C.

IG II² 1277, lines 1–2:

Ἐπὶ Δημοκλέους ἄρχοντος Μουνιχιῶν-
ος [ἑ]β[δ]όμ[η]ι ἐ[π]ὶ δ[έ]κ[α]· ἀγορᾶι κυρίαι

For a financial transaction on this day, see Mounichion 17 in Appendix I.

Mounichion 18

A meeting of the association of the Eranistai occurred on this day near the end of the second century A.D.

144

MOUNICHION

IG II² 1369, lines 24 ff.:

> Ἄρχων μὲν Ταυρίσκος, ἀτὰρ μὴν Μου-
> νιχιὼν ἦν,
> ὀκτ[ω]καιδεκάτῃ δ' ἔρανον σύναγον
> φίλοι ἄνδρες

The sacrificial calendar of the Salaminioi (*Hesp* 1938, pp. 3-5) designates an offering to Eurysakes for this day.

line 88:

> (Μουνιχιῶνος) (85)
> ὀγδόει ἐπὶ δέκα Εὐρυσάκ[ει]: ὗν: ΔΔΔΔ· ξύλα ἐφ'
> ἱεροῖ(ς) καὶ ἐ̣ὶς τἄλλα ⊢⊢⊢.

For a financial transaction on this day, see Mounichion 18 in Appendix I.

Mounichion 19

Plutarch, *Phoc.* 37.1, states that the cavalrymen held a procession for Zeus on this day: ἦν δὲ ἡμέρα μηνὸς Μουνυχιῶνος ἐνάτη ἐπὶ δέκα καὶ τῷ Διὶ τὴν πομπὴν πέμποντες οἱ ἱππεῖς παρεξήεσαν. Deubner (*Feste*, p. 177) follows Mommsen (*Feste*, p. 466) in associating this procession with the Olympieia. From *IG* II² 1496, lines 80-86, the Olympieia must have occurred between the City Dionysia (Elaphebolion 10-16) and the Bendideia (Thargelion 19):

> ἐγ Διονυσίων τῶν ἐν ἄστε[ι] π[αρὰ]
> βοωνῶν: ⌐HHH⌐|⊢⊢⊢
> ἐξ Ὀλυμπιείων παρὰ [τῶν τοῦ]
> δήμου συλ[λο]γέων: ⌐H[⌐]ΔΔ ⊢
> ἐκ τῆς θυσ[ίας] τῶι Ἑρμῆι τῶι
> Ἡγεμονίωι παρὰ [σ]τρατηγῶν[- - -]
> ἐγ Βενδιδέων παρὰ ἱεροποι[ῶν]

The association of the Olympieia with the cavalry procession described by Plutarch is confirmed by the fact that a cavalry contest is attested for the Olympieia (*IG* II² 3079).

Two meetings of the Ekklesia are attested for this day.

III. THE ATHENIAN CALENDAR

IG II² 672, lines 1-3:   279/8 B.C.

['Επ' 'Αναξικράτους ἄρχον]τος ἐπὶ τῆς ['Ακαμαν]τίδος
δεκάτης πρυτανείας ἕ[ι....⁹....]
[.....¹²..... ἐγραμμ]άτευεν· Μουνιχιῶ[ν]ος ἐνάτει ἐπὶ
δέκα, εἰκοστεῖ τ[ῆς πρυτανεί]-
[ας· ἐκκλησία κυρία]

IG II² 775 (see also *Hesp* 1938, p. 145   241/0 B.C.
and *Hesp* 1959, pp. 174-178), lines 28-31:

Ἐπὶ Λυσιά[δο]υ ἄρχοντος ἐπὶ τῆς Ἐ[ρεχθεῖδος δεκάτης]
πρυτανείας ἧι Ἀριστόμαχος Ἀριστο [.....ᶜᵃ· ¹⁶........ ἐ]-
γραμμάτευεν· Μουνιχιῶνος ἐνάτει ἐπ[ὶ δέκα, ἑβδόμει]
[καὶ εἰκο]-
στεῖ τῆς πρυτανείας· ἐκκλησία κυρία

The two meetings *supra* are evidently instances of meetings occurring on a festival day. For a possible financial transaction on this day, see Mounichion 19 in Appendix I.

Restoration to give Mounichion 19:

A meeting of the Ekklesia in 188/7 B.C., *IG* II² 891, lines 1-3 as restored by Meritt, *AJP* 78 (1957), p. 381. These lines are restored by Kirchner to give Mounichion 11. Meritt later (*Year*, p. 155) restored these lines to give Metageitnion 19.

Mounichion 20

The sacrificial calendar of the deme Erkhia prescribes a sacrifice to Leukaspis on this day.

Column Γ, lines 48-53

[Μ]ονιχιῶνος δ-
εκάτει [πρ]οτ-
έραι, Λευκάσ-
πιδι, Ἐρχιᾶ, ο-
ἶς, νηφάλιος,
οὐ φορά, Δ ⊢⊢

No other indisputable evidence as to the nature of this day survives.

MOUNICHION

Restoration to give Mounichion 20:
A meeting of the Ekklesia in 319/8 B.C., *IG* II² 389, lines 1-6. These lines are restored by Dinsmoor (*Archons*, pp. 18-21) to give Mounichion 30.

Mounichion 21
The sacrificial calendar of the deme Erkhia prescribes a sacrifice to the Tritopatores on this day.

Column Δ, lines 41-46
Μονιχιῶνος δ-
εκάτει ὑστέ-
ραι, Τριτοπα-
τρεῦσι, Ἐρχι,
οἷς, νηφάλιο-
ς, οὐ φορά: Δ⊦⊦

No other indisputable evidence as to the nature of this day survives.

Restoration to give Mounichion 21:
A meeting of the Ekklesia in 208/7 B.C., Dow, *Prytaneis*, pp. 86-88, no. 38, lines 1-3.

Mounichion 22
A meeting in 109/8 B.C. establishes this day as a meeting day for the Ekklesia.

BCH 59 (1935), pp. 64-70 (Y. Béquignon), lines 1-7:

[Ἀθη]ναίων ψήφισμα
[Ἐπὶ Ἰ]άσονος ἄρχοντος τοῦ μετ[ὰ Πολύκλει]-
[τον] ἐπὶ τῆς Αἰγείδος δεκάτης πρυ[τανείας ἧι]
[Ἐπι]φάνης Ἐπιφάνου Λαμπτρεὺς ἐγ[ραμμάτευ]-
[εν· Μ]ουνιχιῶνος ἐνάτη μετ' εἰκάδ[ας, ἐνάτηι]
[κα]ὶ εἰκοστῆι τῆς πρυτανείας· ἐκκ[λησία ἐν τῶι]
[θε]άτρωι

For possible financial transactions on this day, see Mounichion 22 in Appendix I.

147

III. THE ATHENIAN CALENDAR

Restorations to give Mounichion 22:

A meeting of the Ekklesia in 327/6 B.C., *Hesp* 1934, pp. 3-4, no. 5 (Meritt), lines 1-6. See also Meritt, *Year*, pp. 100-101.

A meeting of the Ekklesia in 302/1 B.C., *IG* II^2 502, lines 1-5 as restored by Oliver, *Hesp* 1935, p. 38. Kirchner restored these lines to give Mounichion 28.

Mounichion 23

No indisputable evidence as to the nature of this day survives.

Restoration to give Mounichion 23:

A meeting of the Ekklesia after A.D. 196, *Hesp* 1964, pp. 200-201, no. 52 (Meritt), lines 1-3.

Mounichion 24

No evidence as to the nature of this day survives.

Mounichion 25

Except for a financial transaction (see Mounichion 25 in Appendix I), no indisputable evidence as to the nature of this day survives.

Restorations to give Mounichion 25:

A meeting of the Ekklesia in 334/3 B.C., *IG* II^2 405, lines 1-7, *IG* II^2 414a, lines 1-6, and *IG* II^2 335, lines 2-8 as restored by Schweigert, *Hesp* 1940, pp. 339-340. Kirchner restored *IG* II^2 335 to give Boedromion 25.

Mounichion 26

No evidence as to the nature of this day survives.

Mounichion 27

No evidence as to the nature of this day survives.

Mounichion 28

No indisputable evidence as to the nature of this day survives. For a possible financial transaction, see Mounichion 28 in Appendix I.

Restoration to give Mounichion 28:

A meeting of the Ekklesia in 302/1 B.C., *IG* II^2 502, lines 1–5. Oliver (*Hesp* 1935, p. 38) restored these lines to give Mounichion 22.

Mounichion 29

A meeting in 188/7 B.C. establishes this day as a meeting day for the Ekklesia.

IG II^2 892, lines 1–5:

['Ε]πὶ Συμμάχου ἄρχον[τος ἐπὶ τῆς - - -]-
δος δεκάτης πρυτα[νείας ἧι 'Αρχικλῆς]
Θεοδώρου Θορίκιος ἐγρ[αμμάτευεν· Μου]-
νιχιῶ[νο]ς δευτέραι μετ' [εἰκάδας, ἐνάτει]
καὶ εἰ[κ]οστῆι τῆς πρυτανε[ίας· ἐκκλησία]

A private sacrificial calendar of the first or second century A.D. prescribes a sacrifice to Herakles and the "God" on this day.

IG II^2 1367, lines 26–29:

Μουνιχιῶνος β̄ ἀπιόντος Ἡρ[α]-
κλεῖ καὶ θειῷ ἀλέκτορας β̄, πόπαν[α]
χοίνικος δωδεκόμφαλα ὀρθόνφαλ[α]
ἀνυπερθέτως

Restorations to give Mounichion 29:

A meeting of the Ekklesia in the second century B.C., *IG* II^2 954, lines 1–4.
A meeting of the Ekklesia in the second century B.C., *IG* II^2 955, lines 1–4.

Mounichion 30

A meeting in 293/2 B.C. establishes this day as a meeting day for the Ekklesia.

III. THE ATHENIAN CALENDAR

IG II² 649 as re-edited with a new fragment by Dinsmoor, *Archons*, pp. 3–15, lines 1–4:

['Επ]ὶ 'Ο[λυμ]πι[ο]δώρο[υ] ἄ[ρ]χο[ντ]ος ἀναγραφέως δὲ
'Επ[ικ]-
[ού]ρο[υ το]ῦ 'Επιτέλου[ς] 'Ρ[αμν]ουσίου, ἐπὶ τῆς
Πανδ[ιο]-
[νί]δος δεκ[άτ]ης πρυ[τ]ανε[ία]ς· Μουν[ι]χιῶνος [ἕ]νηι
[κα]-
[ὶ νέ]αι, πρώτ[η]ι [τ]ῆς πρυτα[νε]ίας· ἐκκλησία

Restorations to give Mounichion 30:

A meeting of the Ekklesia in 319/8 B.C., *IG* II² 389, lines 1–6 as restored by Dinsmoor, *Archons*, pp. 18–21. These lines are restored by Kirchner to give Mounichion 20.
A meeting of the Ekklesia in 306/5 B.C., *IG* II² 472, lines 2–6.
A meeting of the Ekklesia in 293/2 B.C., *Hesp* 1938, pp. 97–100, no. 17 (Meritt), lines 1–7.

Mounichion—Summary

The Delphinia (Mounichion 6), the Mounichia (Mounichion 16), and the Olympieia (Mounichion 19) were all one-day festivals and are firmly dated. No other annual festivals can be assigned to Mounichion.

150

THARGELION

Thargelion 1

This day was a monthly festival day—the Noumenia.

Restoration to give Thargelion 1:
A meeting of the Ekklesia in 330/29 B.C., *IG* II² 351 addenda, lines 2–7. The date Thargelion 11 is inscribed on the stone, but Meritt (*Year*, pp. 91–94) assumes a scribal error and then "corrects" these lines to give Thargelion 1.

Thargelion 2

This day was a monthly festival day devoted to the Agathos Daimon. From *IG* II² 373, lines 16–20 it would appear that a meeting of the Ekklesia occurred on this day in 322/1 B.C.:

['Επ]ὶ Φιλοκλέους ἄρχοντος ἐπὶ τῆς Οἰνεῖδος ἐνά[τ]-
[ης] πρυτανέας ἧι Εὐθυγένης Ἡφαιστοδήμου Κηφι[σ]-
[ιε]ὺς ἐγραμμάτευεν· Θαργηλιῶνος δευτέραι ἱστα-
[μέ]νου, τρίτει καὶ εἰκοστεῖ τῆς πρυτανείας· ἐκκ[λ]-
[ησ]ία

But in fact the date of this meeting is quite uncertain. The problem lies in the correspondence of the prytany date with the day date. Thargelion 2 simply cannot be the twenty-third day of the ninth prytany. Meritt (*Year*, pp. 111–112) has revised the prytany date, Dinsmoor (*Archons*, pp. 373–374) has revised Thargelion 2 to Thargelion 12, and Pritchett and Neugebauer (*Calendars*, p. 60) have shown how either the day date or the prytany date must be changed in order to create the proper correspondence. The conclusion must be that the day date may well be incorrect, and thus this inscription cannot establish this day as a meeting day for the Ekklesia.

For a financial transaction on this day see Thargelion 2 in Appendix I.

Restorations to give Thargelion 2:
A meeting of the Ekklesia in 307/6 B.C., *IG* II² 455, lines 1–4 as restored by Meritt (*Year*, pp. 177–178), and *IG* II² 460, lines 1–5 as restored by Pritchett and Meritt (*Chronology*, pp. 17–18).

III. THE ATHENIAN CALENDAR

IG II² 455 is restored by Kirchner to give Metageitnion 9, and by Pritchett and Meritt (*Chronology*, p. 20) to give Skirophorion 4. *IG* II² 460 is restored by Kirchner to give Elaphebolion 9, and by Meritt (*Year*, pp. 177-178) to give Skirophorion 3.

Thargelion 3
This day was a monthly festival day devoted to Athena.

Thargelion 4
This day was a monthly festival day devoted to Herakles, Hermes, Aphrodite, and Eros. Hermes is listed among the recipients of sacrifices prescribed for this day on the sacrificial calendar of Erkhia.

Column A, lines 52-56
[Θ]αργηλιῶνος τ-
ετράδι ἱσταμ-
ένο, Λητοῖ ἐμ Π-
[υ]θίο Ἐρχιᾶσι,
[α]ἴξ: Δ

Column Β, lines 45-54
[Θ]αργηλιῶνος
τετράδι ἱστ-
αμένο, Ἀπόλλ-
ωνι Πυθίωι, Ἐ-
ρχι: αἴξ, παρα-
δόσιμος Πυθ-
αισταῖς, Δ ⊦⊦
Ἀπόλλωνι Πα-
ιῶνι, ἐμ Πάγω-
ι Ἐρχι, οἶς, Δ ⊦⊦

Column Γ, lines 54-58
[Θ]αργηλιῶνος
τετράδι ἱστ-
αμένο, Διί, ἐμ
Πάγωι Ἐρχιᾶ,
οἶς, Δ ⊦⊦

Column Ε, lines 47-58
Θαργηλιῶνο-
ς τετράδι ἱ-
σταμένο, Ἑρ-
μῆι, ἐν ἀγορ-
ᾶι Ἐρχιᾶσι,
κριός, τούτ-
ωι ἱερεώσθ-
αι τὸν κήρυ-
κα καὶ τὰ γέ-
ρα λαμβάνε-
ν καθάπερ ὁ
δήμαρχος, Δ

Column Δ, lines 47-51
Θαργηλιῶνος
τετράδι ἱστ-
αμένο, Ἀνάκο-
ιν, Ἐρχιᾶσιν,
οἶς, Δ ⊦⊦

152

THARGELION

Thargelion 5

No evidence as to the nature of this day survives.

Thargelion 6

This day was a monthly festival day devoted to Artemis. Thargelion 6 is also specified as the day on which the Athenians purified their city: Diogenes Laertios, 2.44, Θαργηλιῶνος ἕκτῃ, ὅτε καθαίρουσιν Ἀθηναῖοι τὴν πόλιν καὶ τὴν Ἄρτεμιν Δήλιοι γενέσθαι φασίν. Deubner (*Feste*, p. 179) associates this purification with the festival of the Thargelia on the following day. His evidence for this (Photios and Suda Θαργήλια, and *Etym. Magn.* 443.20) is not, however, conclusive. But from the purification and from the monthly celebration Thargelion 6 may be considered a festival day. A sacrificial calendar of the fifth century B.C. specifies offerings for this day, but unfortunately the deities to whom the offerings were given are unknown.

IG I² 842, lines 5–9:

[Θαρ]γελι[ο̄ν]ο[ς] hέκ[τει h]-
ισταμέ[ν]ο[..]ι[....⁸....]-
εσι τέλεο[ν κ]αὶ ἀ[.]οχὲς ἔ-
μισυ ἒ το̄ι hέροι καὶ φρύ-
γανα

Thargelion 7

Plutarch's discussion of the birthday of Plato establishes this day as the day of the Thargelia.

Plutarch, *Mor.* 717B: τῇ ἕκτῃ τοῦ Θαργηλιῶνος ἱσταμένου τὴν Σωκράτους ἀγαγόντες γενέθλιον τῇ ἑβδόμῃ τὴν Πλάτωνος ἤγομεν.

Plutarch, *Mor.* 717D: ὁ δὲ Φλῶρος οὐδὲ Καρνεάδην ἀπαξιοῦν ἔφη μνήμης ἐν τοῖς Πλάτωνος γενεθλίοις, ἄνδρα τῆς Ἀκαδημίας εὐκλεέστατον ὀργιαστήν· Ἀπόλλωνος γὰρ ἀμφοτέρους ἑορτῇ γενέσθαι, τὸν μὲν [γὰρ] Θαργηλίοις Ἀθήνησιν, τὸν δὲ Κάρνεια Κυρηναίων ἀγόντων· ἑβδόμῃ δ' ἀμφοτέρας ἑορτάζουσιν. . . .

III. THE ATHENIAN CALENDAR

The month of the Thargelia is established by the name of the month Thargelion, which was named after the festival.

Thargelion 8

This day was a monthly festival day devoted to Poseidon and Theseus.

Thargelion 9

No evidence as to the nature of this day survives.

Thargelion 10

A meeting in 218/7 (?) B.C. establishes this day as a meeting day for the Ekklesia.

IG II² 843, lines 2-6:

['Επὶ ....¹⁰.... ἄρχ]οντος ἐπὶ τῆς Αἰγεῖδος δωδε-
[κάτης πρυτανείας ἧ]ι 'Αριστοτέλης Θεαινέτου Κε-
[φαλῆθεν ἐγραμμάτε]υεν· Θαργηλιῶνος δεκάτει
[ἱσταμένου, ἕκτει τῆς] πρυτανείας· ἐκκλησία κυρί-
[α]

Thargelion 11

Four meetings establish this day as a meeting day for the Ekklesia.

IG VII 4253, lines 2-7:   332/1 B.C.

'Επὶ Νικήτου ἄρχοντος ἐπὶ τῆς 'Ερεχθη-
ίδος ἐνάτης πρυτανείας ἧι 'Αριστόνο-
υς 'Αριστόνου 'Αναγυράσιος ἐγραμμάτ-
ευεν· Θαργηλιῶνος ἑνδεκάτει, τρίτηι
καὶ εἰκοστῆι τῆς πρυτανείας· ἐκκλησία κυρία

IG VII 4252, lines 2-7:   332/1 B.C.

'Επὶ Νικήτου ἄρχοντος ἐπὶ τῆς 'Ερε-
χθηίδος ἐνάτης πρυτανέας ἧι 'Αρι-
στόνους 'Αριστόνου 'Αναγυράσιος

ἐγραμμάτευεν· ἑνδεκάτηι, τρίτηι
καὶ εἰκοστῆι τῆς πρυτανείας· ἐκκλησία κυρία

IG VII 4253, line 5 (*supra*) establishes that the month is Thargelion.

IG II² 351 addenda, lines 2-7: 330/29 B.C.

['Επὶ 'Αριστ]οφῶντος ἄρχοντο[ς]
[ἐ]π[ὶ τῆς] Λεωντίδος ἐνάτη[ς] π[ρυ]-
ταν[εία]ς ἧι 'Αντίδωρος 'Αντί[νου]
Παι[ανι]εὺς ἐγραμμάτευεν· ἑ[ν]-
[δ]εκά[τ]ηι Θαργηλιῶνος, ἐνάτη[ι]
[κ]αὶ δ[ε]κάτηι τῆς πρυτανείας

The prytany date is evidently incorrect (see Pritchett and Neugebauer, *Calendars*, p. 50). Meritt (*Year*, pp. 91-94) assumes that the prytany date is correct, and then revises the day date to give Thargelion 1.

IG II² 770, lines 1-4: 262/1 B.C.

['Ε]πὶ Κλεομάχου ἄρχοντος ἐπὶ τῆς 'Αντιοχίδος
ἑνδεκάτη[ς πρυ]-
τανείας ἧι 'Α[φ]θόνητος 'Αρχίνου Κήττιος
ἐγραμμάτευεν· Θ[αρ]-
[γ]ηλιῶνος ἑνδεκάτει, ἑνδεκάτει τῆς πρυτανείας·
ἐκκλ[ησί]-
[α κυρ]ία

Hesp 1964, pp. 183-184, no. 34 beginning of the
(Meritt), lines 1-4: second century B.C.

['Επὶ _ca. 8_ ἄρχοντος ἐπὶ τῆ]ς ['Ι]ππο[θωντί]δος
ἐν[δ]εκάτης πρυταν[εί]-
[ας ἧι _ _ _ _ca. 19_ _ _ _ ε]ὺς ἐγ[ραμ]μάτευεν·
Θαρ[γ]ηλιῶνος ἑνδ[ε]-
[κάτει κατ' ἄρχοντα, κατὰ θεὸν] δὲ ὀγδόει ἐπὶ δέκα,
τρίτει καὶ εἰκοστ[ε͂ι]
[τῆς πρυτανείας· ἐκκλησία κυ]ρία ἐν τῶ[ι] θε[ά]τρωι

For a financial transaction on this day, see Thargelion 11 in Appendix I.

155

III. THE ATHENIAN CALENDAR

Thargelion 12

No evidence as to the nature of this day survives.

Thargelion 13

No indisputable evidence as to the nature of this day survives.

Restoration to give Thargelion 13:

A meeting of the Ekklesia in 324/3 B.C., *IG* II² 362, lines 1-6. Dinsmoor (*Archons*, pp. 372-373) restores these lines to give Thargelion 18. See also *Hesp* 1941, p. 47.

Thargelion 14

Three meetings establish this day as a meeting day.

Aristotle, *Ath. Pol.* 32: 412/1 B.C.

ἡ μὲν βουλὴ [ἡ] ἐπὶ Καλλίου πρὶν διαβουλεῦσαι κατελύθη μηνὸς Θαργηλιῶνος τετράδι ἐπὶ δέκα

IG II² 352, lines 2-10: 330/29 B.C.

['Επὶ 'Αριστ]οφῶντ[ος ἄρχον]-
[τος ἐπὶ τ]ῆς Λεων[τίδος ἐν]-
[άτης πρυ]τανεία[ς ἧι 'Αντί]-
[δωρος 'Αν]τίνου Πα[ιανιεὺ]-
[ς ἐγραμμ]άτευεν· Θα[ργηλι]-
[ῶνος τε]τράδι ἐπὶ δ[έκα, δε]-
[υτέραι] καὶ τριακοσ[τῆι τ]-
[ῆς πρυτα]νείας· ἐκκλη[σία]
[κυρία]

IG II² 790, lines 2-5: 235/4 B.C.

Ἐπὶ Λυσανίου ἄρχοντος ἐπὶ τῆς Αἰαντίδος ἑνδεκάτης π-
ρυτανείας ἧι Εὔμηλος Ἐμπεδίωνος Εὐωνυμεὺς ἐγραμμά-
τευεν· Θαργηλιῶνος τετράδι ἐπὶ δέκα, τετάρτηι καὶ δεκάτηι τῆς πρυτανείας· ἐκκλησία

Thargelion 15

No evidence as to the nature of this day survives.

THARGELION

Thargelion 16

A meeting in 107/6 B.C. establishes this day as a meeting day for the Ekklesia.

IG II² 1011, lines 73-74:

['Αγα]θῇ τύχῃ· Ἐπὶ Ἀριστάρχου ἄρχοντος ἐπὶ τῆς Κεκροπίδος ἑνδεκάτης πρυτανείας ᾗ Τελέστης Μηδείου Παιανιεὺς ἐγραμμάτευεν· Θαργηλιῶνος ἕκτῃ [ἐπὶ] δέκα, ἕκτῃ καὶ δεκάτῃ τῆς πρυτανείας· ἐκκλησία ἐν τῷ θεάτρωι ἐν Διονύσου

The sacrificial calendar of Erkhia prescribes a sacrifice to Zeus Epakrios on this day.

Column E, lines 59-64
(Θαργηλιῶνος) (47)
ἕκτηι ἐπὶ δ-
έκα, Διὶ Ἐπ[α]-
κρίωι, ἐν Ὑμ-
ηττῶι, ἀρ[ήν,]
νηφάλι[ος, ο]-
ὐ φορά, [⊓⊢⊢]

Thargelion 17

No evidence as to the nature of this day survives.

Thargelion 18

A meeting in 302/1 B.C. establishes this day as a meeting day for the Ekklesia.

IG II² 503, lines 1-5:

Ἐπὶ Νικοκλέους ἄρχοντ[ος] ἐπὶ τῆς Ἀ[ντιοχί]-
δος ἑνδεκάτης πρυτανε[ί]ας ἧι Νίκ[ων Θεοδώ]-
ρου Πλωθεεὺς ἐγραμμάτε[υ]εν· Θαρ[γηλιῶνος]
ὀγδόει ἐπὶ δέκα, ἐνάτει κ[α]ὶ δεκά[τει τῆς πρ]-
υτανείας· ἐκκλησία κυρία

III. THE ATHENIAN CALENDAR

Restorations to give Thargelion 18:

A meeting of the Ekklesia in 324/3 B.C., *IG* II² 362, lines 1-6 as restored by Dinsmoor, *Archons*, pp. 372-373. See also *Hesp* 1941, p. 47. Kirchner restores these lines to give Thargelion 13.
A meeting of the Ekklesia in 305/4 B.C., *Hesp* 1936, pp. 201-203 (Meritt), lines 1-7.

Thargelion 19

This day was a festival day—the day of the Bendideia. Commentators on Plato establish the date:

Proklos *in Ti.* 9B: ὅτι γὰρ τὰ ἐν Πειραιεῖ Βενδίδια
τῇ ἐνάτῃ ἐπὶ δέκα τοῦ Θαργηλιῶνος, ὁμολογοῦσιν οἱ περὶ
τῶν ἑορτῶν γράψαντες

Scholion to *Rep.* 327A: ταῦτα [Βενδίδεια] δὲ ἐτελεῖτο
Θαργηλιῶνος ἐνάτῃ ἐπὶ δέκα

Proklos *in Ti.* 27A: Ἀριστοκλῆς ὁ Ῥόδιος ἱστορεῖ τὰ
μὲν ἐν Πειραιεῖ Βενδίδεια τῇ εἰκάδι τοῦ Θαργηλιῶνος
ἐπιτελεῖσθαι

Because the festival extended into the evening of Thargelion 19, according to ancient methods of time-reckoning it technically included two days, Thargelion 19 and 20. For this reason Proklos gives the two different dates for the festival.[56]

The sacrificial calendar of Erkhia prescribes a sacrifice to Menedeios on this day.

Column Δ, lines 52-55
(Θαργηλιῶνος)　　　　　　　　　　(47)
ἐνάτει ἐπὶ δ-
έκα, Μενεδεί-
ωι, Ἐρχιᾶσιν,
οἷς, οὐ φορ, Δ ⊢⊢

Jameson (*BCH* 89, 1965, pp. 158-159) suggests a possible association of this offering to Menedeios with the Bendideia in Athens.

[56] *Ibid.*, p. 488, and Deubner, *Feste*, p. 219.

THARGELION

Restoration to give Thargelion 19:
A meeting of the Ekklesia in 190/89 B.C., Pritchett and Meritt, *Chronology*, pp. 123–126, lines 1–4.

Thargelion 20
No evidence as to the nature of this day survives.

Thargelion 21
No indisputable evidence as to the nature of this day survives.

Restoration to give Thargelion 21:
A meeting of the Ekklesia in 229/8 B.C., *IG* II^2 833, lines 1–4.

Thargelion 22
On this day in 411 B.C. the Four Hundred entered office: Aristotle, *Ath. Pol.* 32, οἱ δὲ τετρακόσιοι εἰσῇεσαν ἐνάτῃ φθίνοντος Θαργηλιῶνος. This does not establish that a meeting was held on this day, but rather that their term of office began on this day. No other evidence concerning the nature of this day survives.

Restoration to give Thargelion 22:
A meeting of the Ekklesia in 299/8 B.C., *IG* II^2 642, lines 2–9.

Thargelion 23
Three meetings establish this day as a meeting day.

IG II^2 1673, lines 9–10   ca. 327/6 B.C.
   [κατὰ ψήφι]σμα τοῦ δήμου ὃ Χαρικλείδης εἶπεν,
    Θαργηλιῶνος μηνὸς ὀγδόηι φθίν[ον]-
   [τος]

IG II^2 361 as revised by D. M. Lewis,   325/4 B.C.
BSA 49 (1954), p. 50, lines 1–5:
   ['Επὶ 'Αντι]κλείους ἄρχοντος ἐπὶ τῆς 'Ακαμαν[τ]-
   [ίδος δ]εκάτης πρυτανείας ἧι 'Αντιφ[ῶν]
   [Κορ]οίβου 'Ελευσίνιος ἐγραμμάτευεν· Θ[αργηλι]-
   [ῶ]νος ὀγδόηι μετ' εἰκάδας, πέμπτ[ηι τῆς πρυ]-
   [ταν]είας· βουλὴ ἐν βουλευτηρίωι

III. THE ATHENIAN CALENDAR

Hesp 1941, pp. 275-277, no. 73   196/5 B.C.
(Pritchett), lines 1-4:

Ἐπὶ Χαρικλέους ἄρχον[τος ἐπὶ τῆς ...⁸.... ἑνδεκάτης]
[πρυτανείας]
ἧι Αἰσχρ(ί)ων Εὐαινέτου ʿΡα[μνούσιος ἐγραμμάτευεν·]
[δήμου ψηφίσματα·]
Θαργηλιῶνος ὀγδόει με[τ' εἰκάδας, ὀγδόει καὶ εἰκοστεῖ]
[τῆς πρυτανείας·]
ἐκκλησία κυρία ἐν τῶι θ[εάτρωι]

Restoration to give Thargelion 23:

A meeting of the Boule in 163/2 B.C., *IG* II² 783, lines 1-4. Meritt (*Hesp* 1934, pp. 29-31) restores these lines to give Skirophorion 23.

Thargelion 24

No indisputable evidence as to the nature of this day survives, but refer also to Thargelion—Summary.

For a possible financial transaction on this day, see Thargelion 24 in Appendix I.

Thargelion 25

Plutarch, *Alc.* 34.1, establishes this day as the day of the Plynteria: ἐδρᾶτο τὰ Πλυντήρια τῇ θεῷ. δρῶσι δὲ τὰ ὄργια Πραξιεργίδαι Θαργηλιῶνος ἕκτῃ φθίνοντος ἀπόρρητα. The month and a date in the last third of the month are confirmed by Proklos, *in Ti.* 27A: Ἀριστοκλῆς ὁ ʿΡόδιος ἱστορεῖ τὰ μὲν ἐν Πειραιεῖ Βενδίδεια τῇ εἰκάδι τοῦ Θαργηλιῶνος ἐπιτελεῖσθαι, ἕπεσθαι δὲ τὰς περὶ τὴν Ἀθηνᾶν ἑορτάς. According to Deubner (*Feste*, p. 18) τὰς περὶ τὴν Ἀθηνᾶν ἑορτάς must refer to the Plynteria and Kallynteria.

Photios, however, dates the Plynteria to Thargelion 29: Καλλυντήρια καὶ Πλυντήρια· γίνονται μὲν αὗται Θαργηλιῶνος μηνός ... δευτέρᾳ δὲ φθίνοντος τὰ Πλυντήρια. The Plynteria was an ἡμέρα ἀποφράς,[57] and it is very improbable that any public meetings could be held on such a day.[58] There are, however, three

---

[57] Plutarch, *Alc.* 34.1, and Xenophon, *Hell.* 1.4.12.
[58] Deubner, *Feste*, p. 18.

meetings of the Ekklesia attested for Thargelion 29 (see Thargelion 29). One may therefore reject the date given by Photios and accept the testimony of Plutarch.

Thargelion 26

No indisputable evidence as to the nature of this day survives.

Restorations to give Thargelion 26:

A meeting of the Ekklesia in 304/3 B.C., *IG* II^2 485, lines 1-5. See also *Hesp* 1937, pp. 323-327.

The taking of an oath in 167/6 B.C., *IG* II^2 951 addenda, lines 1-2. Kirchner earlier (*IG* II^2 951) restored these lines to give Posideon 26.

Thargelion 27

No indisputable evidence as to the nature of this day survives.

Restoration to give Thargelion 27:

A meeting of the Ekklesia in 338/7 B.C., *IG* II^2 237 addenda, lines 1-4. Kirchner earlier (*IG* II^2 237) restored these lines to give Thargelion 29. See also Meritt, *Year*, pp. 73-76.

Thargelion 28

No indisputable evidence as to the nature of this day survives.

Restoration to give Thargelion 28:

A meeting of the Ekklesia in 137/6 B.C., *IG* II^2 974, lines 1-4 as restored by Meritt, *Year*, p. 189. These lines are restored by Kirchner to give Gamelion 28. See also *Hesp* 1959, pp. 188-194.

Thargelion 29

Three meetings establish this day as a meeting day for the Ekklesia.

Aeschines 3.27:   338/7 B.C.

ἐπὶ γὰρ Χαιρώνδου ἄρχοντος, Θαργηλιῶνος μηνὸς
δευτέρᾳ φθίνοντος, ἐκκλησίας οὔσης ἔγραψε Δημοσθένης

III. THE ATHENIAN CALENDAR

IG II² 585 as restored by Meritt,   300/299 B.C.
Hesp 1963, p. 4, lines 1-7:

['Εφ' Ἡγεμάχου ἄρχοντος ἐπὶ τῆς]
[... ντίδος ἑνδεκάτης πρ]υτα[ν]-
[είας ἧι .....¹²....] ϙάνδρου
[....⁹.... ἐγραμμ]άτευεν· Θαρ-
[γηλιῶνος δευτέρ]αι μετ' εἰκάδ-
[ας, τριακοστἔι τῆς] πρυτανείας·
[ἐκκλησία]

Inscriptions de Délos 1505,   146/5 B.C.
lines 41-43:

Ἐπὶ Ἐπικράτου ἄρχοντος ἐπὶ τῆς Ἀτταλίδος
   ἑνδεκάτης
πρυτανείας· Θαργελιῶνος [δ]ευτέραι μετ' εἰκάδας,
   ἐνάτει καὶ
[ε]ἰκοστἔι τῆς πρυτανείας, [ἐκκλη]σία ἐμ Πειραεῖ

Restorations to give Thargelion 29:

Sacrifices on the Athenian State Calendar, *Hesp* 1935, p. 21 (Oliver), Column I, lines 5-17.
A meeting of the Ekklesia in 338/7 B.C., *IG* II² 237, lines 1-4. Kirchner later (*IG* II² 237 addenda) restored these lines to give Thargelion 27. See also Meritt, *Year*, pp. 73-76.
A meeting of the Ekklesia in 324/3 B.C., *IG* II² 547, lines 1-6 as restored by Pritchett and Meritt, *Chronology*, pp. 2-3. Kirchner restored these lines to give Pyanopsion 29.
A meeting of the Ekklesia in 320/19 B.C., *IG* II² 383b addenda, lines 3-8. Pritchett and Neugebauer (*Calendars*, p. 62) accept this restoration to give Thargelion 29. Meritt (*Year*, pp. 113-114) restores these lines to give Boedromion 29.

Thargelion 30

No indisputable evidence as to the nature of this day survives. *IG* II² 375, lines 1-7 would appear to attest a meeting of the Ekklesia on this day in 322/1 B.C.

THARGELION

['Επὶ Φι]λοκ[λέους ἄρχοντος ἐπὶ τ]-
[ῆς Πανδι]ονίδος δεκάτης [πρυτα]-
[νεία]ς ἧι Εὐθυγένης Ἡφαισ[τοδή]-
[μο]υ Κ[ηφ]ισιεὺς ἐγρα[μ]μάτε[υεν· Θ]-
[αρ]γηλιῶνος ἕνηι καὶ ν[έ]αι, [....]
[... κ]αὶ τριακοστῆι [τ]ῆ[ς] πρ[υταν]-
[είας· ἐκκ]λ[η]σί[α]

All modern commentators, however, conclude from the prytany number and day that Thargelion was mistakenly inscribed for Skirophorion (Dinsmoor, *Archons*, pp. 373-374; Pritchett and Neugebauer, *Calendars*, p. 60; Meritt, *Year*, pp. 111-112).

Restorations to give Thargelion 30:
A meeting of the Ekklesia in 327/6 B.C., *IG* II^2 357, lines 2-6 as restored by Pritchett and Neugebauer, *Calendars*, p. 53. Kirchner restores these lines to give Posideon 30.
A meeting of the Ekklesia in 281/0 B.C., *IG* II^2 660, lines 25-27.

Thargelion—Summary

The Kallynteria occurred in Thargelion, but the specific day of the festival is uncertain. Photios gives Thargelion 19 as the day of the festival: Καλλυντήρια καὶ Πλυντήρια· ἑορτῶν ὀνόματα. γίνονται μὲν αὗται Θαργηλιῶνος μηνός, ἐννάτῃ μὲν ἐπὶ δέκα Καλλυντήρια, δευτέρᾳ δὲ φθίνοντος τὰ Πλυντήρια. There are, however, three considerations which suggest that this citation in Photios is inaccurate. In the first place, the date of Thargelion 29 for the Plynteria is incorrect (see Thargelion 25). Secondly, it is very improbable that the Bendideia, which did occur on Thargelion 19 (see Thargelion 19), could have displaced the Kallynteria or could have been celebrated on the same day.[59] And, finally, Proklos indicates that both the Kallynteria and the Plynteria followed the Bendideia: in *Ti.* 27A, Ἀριστοκλῆς ὁ Ῥόδιος ἱστορεῖ τὰ μὲν ἐν Πειραιεῖ Βενδίδεια τῇ εἰκάδι τοῦ Θαργηλιῶνος ἐπιτελεῖσθαι, ἕπεσθαι δὲ τὰς περὶ τὴν Ἀθηνᾶν ἑορτάς.

If we reject Photios' dating of the Kallynteria, we must date the Kallynteria in the last third of Thargelion (see Proklos, *in Ti.* 27A

[59] *Ibid.*

III. THE ATHENIAN CALENDAR

cited *supra*). Deubner (*Feste*, pp. 17-18) notes that the Kallynteria and Plynteria are of similar character, and that they are often associated in the sources.[60] For this reason he dates the Kallynteria to Thargelion 24, the day preceding the Plynteria. This is certainly a possibility, but there is no conclusive evidence that the Kallynteria had to precede the Plynteria. From the calendric study of Thargelion *supra* it is clear that no meetings are attested for the period Thargelion 24-28. It would seem that the Kallynteria could be dated to any one of these days, with the exception of Thargelion 25, the day of the Plynteria.

[60] Photios Καλλυντήρια καὶ Πλυντήρια, and *Etym. Magn.* 487.13.

SKIROPHORION

Skirophorion 1

This day was a monthly festival day—the Noumenia.

Skirophorion 2

This day was a monthly festival day devoted to the Agathos Daimon. A meeting of the phylai occurred on this day in 338/7 B.C.

Aeschines 3.27:

ἔγραψε Δημοσθένης ἀγορὰν ποιῆσαι τῶν φυλῶν
Σκιροφοριῶνος δευτέρᾳ ἱσταμένου καὶ τρίτῃ

The association of the Thiasotai of Bendis on Salamis regularly met on this day.

IG II² 1317, line 1:   272/1 B.C.

Ἐπὶ Λυ[σ]ιθείδου ἄρχοντος Σκιροφοριῶνος δευτέρα[ι·]
[ἀγορᾶι κυρίαι]

SEG 2, no. 10, lines 2–3:   250/49 B.C.

Ἐπὶ Θερσιλόχου ἄρχοντος, Σκιροφοριῶνος δευτέραι
ἱσ-
ταμένου, κυρίαι ἀγορᾶι

IG II² 1317b addenda, lines 2–3:   248/7 B.C.

Ἐπὶ Ἱέρωνος ἄρχοντος καὶ ἱερέως Χαρίνου· μηνὸς
Σκιροφοριῶνος
δευτέραι ἱσταμένου· κυρίαι ἀγορᾶι

Legal proceedings are recorded for this day in 342/1 (?) B.C.

Hesp 1936, pp. 393–413, no. 10
(Meritt), lines 115–116:

Σκιροφοριῶ[νος δε]-
υτέραι ἱσταμένου δικαστήριον

Restorations to give Skirophorion 2:

Sacrifices on the Athenian State Calendar, *Hesp* 1935, p. 21 (Oliver), lines 19–27 as restored (unpublished) by Dow.

III. THE ATHENIAN CALENDAR

A meeting of the Ekklesia in 339/8 B.C., *Hesp* 1938, pp. 291-292, no. 18 (Schweigert), lines 1-3. See also *BSA* 51 (1956), pp. 2-3.

Skirophorion 3

This day was a monthly festival day devoted to Athena. A meeting of the phylai and a meeting of the Boule are also attested for this day.

Aeschines 3.27:

ἔγραψε Δημοσθένης ἀγορὰν ποιῆσαι τῶν φυλῶν Σκιροφοριῶνος δευτέρᾳ ἱσταμένου καὶ τρίτῃ

IG II² 847, lines 1-6:   215/4 B.C.

Ἐπὶ Διοκλέους ἄρχοντο(ς) ἐπὶ τῆς Ἱπποθωντίδος τρίτης καὶ δεκάτης πρυτανείας ἧι Ἀριστοφάνης Στρατοκλέους Κειριάδης ἐγραμμάτευεν· βουλῆς ψηφίσματα· Σκιροφοριῶνος τρίτει ἱσταμένου, τρίτει τῆς πρυτανείας· βουλὴ ἐμ βουλευτηρίωι

The sacrificial calendar of Erkhia prescribes sacrifices for this day to Kourotrophos, Athena Polias, Aglauros, Zeus Polieus, and Poseidon.

Column A, lines 57-65
[Σ]κιροφοριῶνο-
ς τρίτη ἱσταμ-
ένου Κουροτρ-
όφωι, ἐμ Πόλει
Ἐρχ: χοῖρος, ⊢⊢⊢
Ἀθηνάαι Πολι-
άδι, ἐμ Πόλει Ἐ-
ρχιᾶσι, οἷς, ἀν-
[τ]ίβους, Δ

Column B, lines 55-59
[Σ]κιροφοριῶν-
ος τρίτηι ἱσ-
ταμένο, Ἀγλα-
ύρωι, ἐμ Πόλε
Ἐρχι: οἷς, Δ

Column Γ, lines 59-64
[Σ]κιροφοριῶν-
ος τρίτει ἱσ-
ταμένο, Διὶ Π-
ολιεῖ, ἐμ Πόλε
Ἐρχιᾶσι, οἷς,
οὐ φορά, Δ⊢⊢

Column Δ, lines 56-60
Σκιροφοριῶν-
ος τρίτηι ἱσ-
ταμένο, Ποσε-
ιδῶνι, ἐμ Πόλ-
ε Ἐρχι: οἷς: Δ⊢⊢

SKIROPHORION

M. Jameson (*BCH* 89, 1965, pp. 156–157) and W. Burkert (*Hermes* 94, 1966, p. 5, note 2) associate these sacrifices from the Erkhia Calendar with the myths and rites of the Arrephoria. The Arrephoria are attested to have been celebrated in Athens in Skirophorion (*Etym. Magn.* 149.13 Ἀρρηφόροι καὶ Ἀρρηφορία· ἑορτὴ ἐπιτελουμένη τῇ Ἀθηνᾷ ἐν Σκιρροφοριῶνι μηνί). The recipients of the sacrifices (Kourotrophos, Athena Polias, Aglauros, Zeus Polieus, and Poseidon) are the basis for the association with the Arrephoria. These sacrifices are to be presented ἐμ Πόλε Ἐρχιᾶσι, and Jameson suggests that the state festival in Athens may also have occurred on Skirophorion 3, "or thereabouts." Burkert is more cautious, noting that one need not assume that the cults of a deme and of the state correspond exactly. The evidence does suggest strongly that the state Arrephoria are to be dated to Skirophorion 3, but it is not conclusive.

Restorations to give Skirophorion 3:

A sacrifice to Pandrosos on the sacrificial calendar of Erkhia, Column E, lines 65–70, as restored by Jameson, *BCH* 89 (1965), p. 157.

A meeting of the Ekklesia in 307/6 B.C., *IG* II[2] 460, lines 1–5 as restored by Meritt, *Year*, pp. 177–178. These lines are restored by Kirchner to give Elaphebolion 9, and by Pritchett and Meritt (*Chronology*, pp. 17–18) to give Thargelion 2.

Skirophorion 4

This day was a monthly festival day devoted to Herakles, Hermes, Aphrodite, and Eros. A meeting of the Boule is attested for this day in 161/0 B.C.

IG II[2] 952, lines 1–3:

[Ἐπὶ Ἀριστό]λα ἄρχοντος ἐπὶ τῆς Ἱπποθωντίδος
δωδεκάτης πρυτανείας
[ἕι .. $^{ca. 7}$..]ς Φιλωνίδου Ἐλευσίνιος ἐγ[ρ]αμμάτευεν·
Σκιροφοριῶνος τετρά-
[δι ἱσταμέν]ου, τετάρτει τῆς πρυτανε[ίας· βουλὴ ἐμ]
βουλευτηρίωι

III. THE ATHENIAN CALENDAR

Restoration to give Skirophorion 4:

A meeting of the Ekklesia in 307/6 B.C., *IG* II² 455, lines 1–4 as restored by Pritchett and Meritt (*Chronology*, p. 20). Kirchner restores these lines to give Metageitnion 9, and Meritt (*Year*, pp. 177–178) restores these lines to give Thargelion 2.

Skirophorion 5

Except for financial transactions (see Skirophorion 5 in Appendix I), no evidence as to the nature of this day survives.

Skirophorion 6

This day was a monthly festival day devoted to Artemis. A meeting of the Nomothetai dealing with sacred matters is attested for this day in 335/4 B.C.

IG II² 333, line 13:

['Επὶ Εὐαινέτου ἄρχοντος ἐπὶ τῆς ..⁵.. ίδος]
[δεκά]της· Σκιροφορ[ιῶνος ἕκ]τηι ἱσταμένου·
νομο[θετῶν ἕδρα]

This inscription is stoichedon in form, and the spacing requires the restoration to give Skirophorion 6.

Restoration to give Skirophorion 6:

A meeting of the Ekklesia in 303/2 B.C., *IG* II² 498 + addenda, lines 1–6.

Skirophorion 7

This day was a monthly festival day devoted to Apollo.

Skirophorion 8

This day was a monthly festival devoted to Poseidon and Theseus. A meeting of the Thracian orgeones of Bendis is attested for this day in 257/6 (?) B.C.

SKIROPHORION

IG II² 1284, lines 19-21:

Ἐπὶ Λυκέου ἄρχοντος μηνὸς Σ[κιροφο]-
ριῶνος ὀγδόει ἱσταμένου· ἀγο[ρᾶι κυρί]-
αι

Restoration to give Skirophorion 8:
A meeting of the Ekklesia in 267/6 B.C., *IG* II² 664 + addenda, lines 1-5.

Skirophorion 9

A meeting in 240/39 B.C. establishes this day as a meeting day for the Ekklesia.

IG II² 784, lines 1-4:

Ἐπὶ Ἀθηνοδώρου ἄρχοντος ἐπὶ τῆς Ἀντιγονίδος δωδεκάτη-
ς πρυτανείας ἧι Ἄρκετος Ἀρχίου Ἁμαξαντεὺς ἐγραμμάτευ-
εν· Σκιροφοριῶνος ἐνάτει ἱσταμένου, ἐνάτει τῆς πρυτανε-
ίας· ἐκκλησία κυρία

Skirophorion 10

A meeting in 331/0 B.C. establishes this day as a meeting day for the Ekklesia.

IG II² 349, lines 3-7:

Ἐπὶ Ἀριστοφάνους ἄρχοντος
ἐπὶ τῆς Κε[κ]ροπίδος δεκάτης πρυτα-
νείας· Σκ[ιρ]οφοριῶνος δεκάτηι ἱσ[τ]-
αμένου, [ἕκτ]ει καὶ δεκάτει τῆς πρυ[τ]-
ανεία[ς]

Skirophorion 11

A meeting in 272/1 B.C. establishes this day as a meeting day for the Ekklesia.

III. THE ATHENIAN CALENDAR

Hesp 1957, pp. 54–55, no. 11 (Meritt), lines 2–7:

Ἐπὶ Λυσιθείδου ἄρχοντος ἐπὶ τῆς
Κεκροπίδος δωδεκάτης πρυτανείας ἕι Σημωνίδης Τιμησίου Σουνιε(ὺς)
ἐγραμμάτευε· Σκιροφοριῶνος ἐν-
δεκάτει, ἑνδεκάτει τῆς πρυτανεί-
[α]ς· ἐκκλησία

Restorations to give Skirophorion 11:

A meeting of the Ekklesia in 280/79 B.C., *IG* II² 670, lines 1–4 as restored by Meritt (*Hesp* 1938, p. 106). Kirchner originally restored these lines to give Elaphebolion 11, but later (*IG* II² 670 addenda) to give Skirophorion 12. Meritt later (*Hesp* 1969, pp. 109–110) restored these lines to give Gamelion 16.

A meeting of the Ekklesia in the third century B.C., *IG* II² 774, lines 1–4 as restored by Meritt, *Hesp* 1935, pp. 551–552. See also *Hesp* 1938, pp. 144–145.

Skirophorion 12

This day was a festival day—the day of the Skira. The scholion to Aristophanes *Eccl.* 18 establishes the date: Σκίρα ἑορτή ἐστι τῆς Σκιράδος Ἀθηνᾶς, Σκιροφοριῶνος ιβ. Plutarch, *Mor.* 350A, indicates that the battle of Mantineia made this day "more holy":

τὴν δὲ δωδεκάτην τοῦ Σκιρροφοριῶνος ἱερωτέραν ἐποίησεν ὁ Μαντινειακὸς ἀγών, ἐν ᾧ τῶν ἄλλων συμμάχων ἐκβιασθέντων καὶ τραπέντων μόνοι τὸ καθ᾽ ἑαυτοὺς νικήσαντες ἔστησαν τρόπαιον ἀπὸ τῶν νικώντων πολεμίων.

The comparative form ἱερωτέραν suggests that the day was an ἱερὰ ἡμέρα before the battle of Mantineia, and thus confirms the testimony of the scholiast to Aristophanes.

Restoration to give Skirophorion 12:

A meeting of the Ekklesia in 280/79 B.C., *IG* II² 670 addenda, lines 1–4. Kirchner earlier (*IG* II² 670) restored these lines to give Elaphebolion 11. These lines were restored by Meritt to give both Skirophorion 11 (*Hesp* 1938, p. 106) and Gamelion 16 (*Hesp* 1969, pp. 109–110).

SKIROPHORION

Skirophorion 13

No evidence as to the nature of this day survives.

Skirophorion 14

This day was a festival day—the day of the Dipolieia. Two ancient sources establish the date:

Scholiast to Aristophanes *Pax* 419: Διπόλεια δὲ ἑορτὴ Ἀθήνησιν, ἐν ᾗ Πολιεῖ Διὶ θύουσι Σκιρροφοριῶνος τετάρτῃ ἐπὶ δέκα.

Etym. Magn. 210.30: Βουφόνια· ἤγετο αὕτη Σκιροφοριῶνος μηνὸς τετάρτῃ ἐπὶ δέκα.

In view of this evidence the confused citation in Bekker's *Anecd.* 1.238.21 (Διιπόλια γάρ ἐστιν ἑορτὴ μὲν Διί, ἣ καὶ δειλία καλεῖται, γίνεται δὲ ἕκτην ἐπὶ δέκα τοῦ Σκληροφοριῶνος μηνός) must be mistaken in dating the "Diipolia" to Skirophorion 16.

According to Aristotle, *Ath. Pol.* 32, a new Boule was to have entered office on this day in 411 B.C.: ἔδει δὲ τὴν εἰληχυῖαν τῷ κυάμῳ βουλὴν εἰσιέναι δ̄ ἐπὶ δέκα Σκιροφοριῶνος. Other instances of the conciliar year beginning in Skirophorion are attested from the fifth century B.C.[61] One cannot, however, infer from this passage that a meeting of the Boule was to have taken place on Skirophorion 14. The meaning is rather that the term of office for the new Boule was to have begun on this day.

Restoration to give Skirophorion 14:

A meeting of the Ekklesia in 319/8 B.C., *IG* II² 390, lines 1–4 as restored by Dinsmoor, *Archons*, pp. 21–22.

Skirophorion 15

On this day in 145/4 B.C. the Boule of the Athenian cleruchs on Delos met:

Inscriptions de Délos 1506, lines 1–3:

Ἐπὶ Μητροφάνου ἄρχοντος, Σκιροφοριῶνος πέμπτει ἐπὶ δέκα, βουλὴ ἐν τῶι ἐκκλησιαστ[η]-ρίωι

[61] Dinsmoor, *Archons*, pp. 323-325.

III. THE ATHENIAN CALENDAR

This is not sufficient evidence to establish this day as a meeting day in Athens.

Skirophorion 16

Three meetings establish this day as a meeting day for the Ekklesia.

Demosthenes 19.58:   346 B.C.

ἡ δ' ἐκκλησία μετὰ ταῦτα ... τῇ ἕκτῃ ἐπὶ δέκα ἐγίγνετο τοῦ Σκιροφοριῶνος.

IG II² 893 (see also *AJP* 78, 1957,   188/7 B.C.
pp. 375–381), lines 2–6:

['Επὶ Συμμάχου ἄρχο]ν[το]ς [ἐ]πὶ τῆς Ἀντιοχίδος δω-
[δεκάτης πρυτανείας ἧι Ἀρ]χικλῆς Θεοδώρου Θορί-
[κιος ἐγραμμάτευεν· Σκιρ]οφοριῶνος ἕκτει ἐπὶ [δ]-
[έκα, ὀγδόει καὶ δεκάτ]ει τῆς πρυτανείας· [ἐκκ]λη-
[σία κυρία ἐν τῶι] θεάτρωι

IG II² 949, lines 1–4:   165/4 B.C.

Ἐπὶ Πέλοπος ἄρχοντος ἐπὶ τῆς Πτολεμαιίδος
   δωδεκάτης πρυτανείας
ἧι Διονυσικλῆς Διονυσίου Ἑκαλῆθεν ἐγραμμάτευεν·
   Σκιροφοριῶνος ἕ-
κτει ἐπὶ δέκα, ἕκτει καὶ δεκάτει τῆς πρυτανείας·
   ἐκκλησία ἐν τῶι θε-
άτρωι

See also *IG* II² 950, lines 1–3.

For a financial transaction on this day, see Skirophorion 16 in Appendix I. Refer also to Skirophorion 16 in Appendix II.

Restorations to give Skirophorion 16:

A meeting of the Ekklesia in 337/6 B.C., *Hesp* 1938, pp. 292–294, no. 19 (Schweigert), lines 1–4.

A meeting of the Boule in 199/8 B.C., *Hesp* 1940, pp. 85–86, no. 16 (Meritt), lines 2–6 as restored by Meritt, *Hesp* 1957, pp. 62–63.

Skirophorion 17

No evidence as to the nature of this day survives.

Skirophorion 18

A meeting in 335/4 B.C. establishes this day as a meeting day for the Ekklesia.

SEG 21, no. 272, lines 2-7:

['Επ' Εὐαιν]έτου ἄρχοντος ἐπὶ [τῆς]
['Αντι]οχίδος δεκάτης πρυτα[νεί]-
[ας ἧι] Πρόξενος Πυλαγόρου ['Αχερ]-
[δούσ]ιος ἐγραμμάτευεν· Σ[κιροφ]-
[ορι]ῶνος ὀγδόηι ἐπὶ δέκα, [τ]ρ[ίτη]-
[ι κ]αὶ εἰκοστῆι τῆς πρυταν[είας]

Restoration to give Skirophorion 18:
A meeting of the Ekklesia in 188/7 B.C., *Hesp* 1946, pp. 144-146, no. 6 (Pritchett), lines 1-4. Meritt (*Year*, p. 156) restores these lines to give Anthesterion 18.

Skirophorion 19

No evidence as to the nature of this day survives.

Skirophorion 20

No indisputable evidence as to the nature of this day survives.

Restorations to give Skirophorion 20:
A meeting of a religious association in 112/1 B.C., *Hesp* 1961, p. 229, no. 28 (Meritt), lines 1-4.
A meeting of a religious association in 111/0 B.C., *Hesp* 1961, pp. 229-230, no. 29 (Meritt), lines 2-4.

Skirophorion 21

Three meetings establish this day as a meeting day for the Ekklesia.

III. THE ATHENIAN CALENDAR

IG II² 493, lines 2-8: 303/2 B.C.

Ἐπὶ Λεωστράτου ἄρχοντος ἐπὶ τῆ-
ς Αἰαντίδος δωδεκάτης πρυτανείας ἧι Διόφαντος Διονυσοδώρου
Φηγούσιος ἐγραμμάτευεν· Σκιροφοριῶνος δεκάτει ὑστέραι, τρίτει καὶ εἰκοστῆι τῆς πρυτανείας·
ἐκκλησία κυρία

IG II² 494, lines 2-8: 303/2 B.C.

Ἐπὶ Λεωστράτ[ου ἄρχοντος ἐπὶ τῆς]
[Α]ἰαντίδος δωδ[εκάτης πρυτανεία]-
ς ἧι Διόφαντος Δ[ιονυσοδώρου Φ]-
ηγούσιος ἐγραμ[μάτευεν· Σκιροφο]-
[ρ]ιῶνος δεκάτε[ι ὑστέραι, τρίτει κα]-
[ὶ] εἰκοστῆι τῆς [πρυτανείας· ἐκκλησ]-
[ί]α κυρία

IG II² 505, lines 1-4: 302/1 B.C.

Ἐπὶ Νικοκλέους ἄρχοντος ἐπὶ τῆς Αἰαντίδος δωδεκάτης πρυτανείας ἕι Νίκων Θεοδώρου Πλωθεὺς [ἐ]-
γραμμάτευεν· Σκιροφοριῶνος δεκάτει ὑστέραι, μ-
[ι]ᾶι καὶ εἰκοστῆι τῆς πρυτανείας· ἐκκλησία

IG II² 676, lines 2-6: 273/2 B.C.

Ἐπὶ Γλα[υκίππου ἄρ]χοντος ἐπὶ τῆς [....⁸....]-
ς δωδεκ[άτης πρυτα]νείας ἧι Εὔθοιν[ος ... κ]-
ρίτου [Μυρρινούσι]ος ἐγραμμάτευεν· [Σκιροφ]-
οριῶ[νος δεκάτει ὑσ]τέραι, τρίτει κα[ὶ εἰκοσ]-
τῆι [τῆς πρυτανείας· ἐκ]κλησία

Restorations to give Skirophorion 21:

A meeting of the Ekklesia in 271/0 B.C., Meritt, *Year*, pp. 194-195, lines 2-5.
A meeting of the Ekklesia in 254/3 B.C., *IG* II² 697, lines 1-5 as restored by Dow, *Hesp* 1963, pp. 352-356. See also *Hesp* 1969, pp. 433-434. Kirchner restored these lines to give Elaphebolion 21.

174

SKIROPHORION

Skirophorion 22

Except for a financial transaction (see Skirophorion 22 in Appendix 1) no evidence as to the nature of this day survives.

Skirophorion 23

Four meetings establish this day as a meeting day.

IG II² 772, lines 1-5:   268/7 B.C.
Ἐπὶ Διογείτονος ἄρχοντος ἐπὶ τῆς Δη-
μητριάδος δωδεκάτης πρυτανείας ἧι
Θεόδοτος Θεοφίλου Κειριάδης ἐγραμ-
μάτευεν· Σκιροφοριῶνος ὀγδόει μετ' ε-
ἰκάδας· ἐκκλησία κυρία

IG II² 889 (see also Meritt, *Year*,   181/0 B.C.
p. 197), lines 1-4:
[Ἐπὶ....] ου ἄρχο[ντο]ς ἐ[πὶ τῆς ..ᶜᵃ⋅⁷.. δος δωδεκάτης]
[πρυτανεία]-
[ς ἧι Θε]οδόσιος Ξενοφά[ν ......ᶜᵃ⋅¹⁵...... ἐγραμμάτευεν·]
[Σκιρο]-
[φ]οριῶνος ὀγδόει μετ' εἰκ[άδας, .ᶜᵃ⋅⁵. ι καὶ εἰκοστῆι]
[τῆς πρυτανεί]-
ας· ἐκκλησία κυρία ἐν τῶι θ[εάτρωι]

IG II² 971, lines 9-10:   140/39 B.C.
Ἐπὶ Ἁγνοθέου ἄρχοντος· Σκιροφοριῶνος ὀγδόει
μετ' [εἰκάδας·]
ἐκκλησία ἐν Πειραιεῖ

IG II² 1046, lines 1-5:   52/1 B.C.
Ἀγαθῆι τύχηι τῆς βουλῆς καὶ τοῦ δήμου τοῦ
Ἀθηναίων· ἐπὶ Λυ-
σάνδρου τοῦ Ἀπολήξιδος ἄρχοντος, ἐπὶ τῆς
Πανδιονίδος
δωδεκάτης πρυτανείας ἧ Γάιος Γαίου Ἁλαιεὺς
ἐγραμμά-
τευεν· Σκιροφοριῶνος ὀγδόηι μετ' ἰκάδας, τρίτηι
καὶ εἰκοστῆ
τῆς πρυτανείας· βουλὴ ἐν βουλευτηρίωι

III. THE ATHENIAN CALENDAR

Restorations to give Skirophorion 23:

A meeting of the Ekklesia in beginning of the second century B.C., *Hesp* 1947, p. 187, no. 93 (Pritchett), lines 1-5.
A meeting of the Boule in 163/2 B.C., *IG* II² 783, lines 1-4 as restored by Meritt, *Hesp* 1934, pp. 29-31. Kirchner restored these lines to give Thargelion 23.

Skirophorion 24

Except for two financial transactions (see Skirophorion 24 in Appendix I), no evidence as to the nature of this day survives.

Skirophorion 25

A meeting in 285/4 B.C. establishes this day as a meeting day for the Ekklesia.

IG II² 654, lines 1-7:

['Ε]πὶ Διοτίμου ἄρχοντος ἐπὶ τῆ[ς]
[Π]αμδιονίδος δ[ω]δεκάτης πρυ[τα]-
νείας ἧι Λυσίσ[τ]ρατος ['Α]ριστο[μ]-
άχου Παιανιεὺ[ς] ἐγραμ[μ]άτευε[ν·]
Σκιροφοριῶνος ἕκτει [μ]ετ' εἰκ[ά]-
δας, πέμπτει καὶ ε(ἰ)κοστ[ε͂]ι τῆς π[ρ]-
υτανείας· ἐκκλησία

Restorations to give Skirophorion 25:

A meeting of the Ekklesia in 324/3 B.C., *IG* II² 454, lines 1-4 as restored by Meritt, *Year*, p. 106. Kirchner restored these lines to give Skirophorion 26.
A meeting of the Ekklesia in 285/4 B.C., *IG* II² 655, lines 1-2.

Skirophorion 26

No indisputable evidence as to the nature of this day survives. For a possible financial transaction on this day, see Skirophorion 26 in Appendix I.

SKIROPHORION

Restorations to give Skirophorion 26:

A meeting of the Ekklesia in 324/3 B.C., *IG* II² 454, lines 1-4. Meritt (*Year*, p. 106) restored these lines to give Skirophorion 25. A meeting of the Ekklesia in 234/3 B.C., Pritchett and Meritt, *Chronology*, pp. 100-101, lines 1-6.

Skirophorion 27

Demosthenes (19.60) attests that a meeting of the Ekklesia occurred on this day in 346 B.C., and thus establishes the day as a meeting day: τῇ τετράδι φθίνοντος [Σκιροφοριῶνος] ἠκκλησιάζετε μὲν τόθ' ὑμεῖς ἐν Πειραιεῖ περὶ τῶν ἐν τοῖς νεωρίοις....

Restoration to give Skirophorion 27:

A meeting of the Boule in 192/1 B.C., *IG* II² 916, lines 2-7 as re-edited with new fragments by Pritchett and Meritt, *Chronology*, pp. 113-116.

Skirophorion 28

No evidence as to the nature of this day survives.

Skirophorion 29

Legal proceedings are attested for this day. From Lysias 26.6, ἡ γὰρ αὔριον ἡμέρα μόνη λοιπὴ τοῦ ἐνιαυτοῦ ἐστιν, ἐν δὲ ταύτῃ τῷ Διὶ τῷ σωτῆρι θυσία γίγνεται, it is evident that the speech must have been given in court on the day preceding the last day of the year. If Skirophorion were a full month, this day would have been Skirophorion 29.

Restoration to give Skirophorion 29:

A meeting of the Ekklesia in 302/1 B.C., *IG* II² 504, lines 1-7.

Skirophorion 30

Ten meetings of the Ekklesia establish this day as a meeting day for the Ekklesia.

III. THE ATHENIAN CALENDAR

IG ii² 415, lines 7–11: ca. 330/29 B.C.

Σκιροφοριῶνος ἕνηι καὶ νέαι, τετά[ρτ]-
ει καὶ τριακοστῆι τῆς πρυτανείας· τ[ῶ]-
[ν] προέδρων ἐπεψήφιζεν Δημήτριος Ἐ[ρ]-
[χι]εύς· ἔδοξεν τῆι βουλῆι καὶ τῶι δήμω-
[ι]

IG ii² 486, lines 3–8: 304/3 B.C.

Ἐπὶ Φερεκλέους ἄρχοντος ἐπὶ τῆς [Αἰαντί]-
δος δωδεκάτης πρυτανείας ἧι Ἐπ[ιχαρῖνο]-
ς Δημοχάρου Γαργήττιος ἐγρα[μμάτευεν· Σ]-
κιροφοριῶνος ἕνηι καὶ ν[έαι προτέραι, ἐν]-
άτει καὶ εἰκοστῆι τῆς π[ρυτανείας· ἐκκλη]-
σία

Hesp 1938, p. 297, no. 22 304/3 B.C.
(Schweigert), lines 3–8:

[Ἐπὶ Φερεκλέους ἄ]ρχοντος ἐπὶ τῆς Αἰ -
[αντίδος δωδεκάτ]ης πρυτανείας ἧι Ἐ-
[πιχαρῖνος Δημο]χάρους Γαργήττιος
[ἐγραμμάτευεν· Σ]κιροφοριῶνος ἕνει
[καὶ νέαι προτέρ]αι, ἐνάτηι καὶ εἰκοσ-
[τῆι τῆς πρυτανεί]ας· ἐκκλησία

IG ii² 495, lines 1–6: 303/2 B.C.

[Ἐπὶ] Λεωστράτου ἄρχοντος ἐπὶ τ[ῆς Αἰ]-
[αντί]δος δωδεκάτης πρυτανείας [ἧι Δ]-
[ιόφ]αντος Διονυσοδώρου Φηγούσιο[ς]
[ἐγρα]μμάτευε· Σκιροφοριῶνος ἕνηι κ-
[αὶ νέ]αι προτέραι, μιᾶι καὶ τριακοστ-
[ῆι] τῆς πρυτανείας· ἐκκλησία

See also *IG* ii² 496 and 497.

IG ii² 659, lines 2–7: 283/2 B.C.

Ἐπ' Εὐθίου ἄρχοντος ἐπὶ τῆς
Αἰαντίδος δωδεκάτης πρυ-
τανείας ἧι Ναυσιμένης
Ναυσικύδου Χολαργεὺς

178

ἐγραμμάτευεν· Σκιροφοριῶ-
νος ἔνηι καὶ νέαι

IG II² 685, lines 1-6:   276/5 B.C.

Ἐπὶ Φιλοκράτου ἄρχοντος ἐπὶ τῆς Δη[μ]-
ητριάδος δωδεκάτης πρυτανείας ἧι Ἡ-
[γ]ήσιππος Ἀριστομάχου Μελιτεὺς ἐγρ-
αμμάτευεν· Σκιροφοριῶνος ἔνε[ι] καὶ ν-
έαι, δευτέραι καὶ τριακοστῆι τῆς πρυ-
τανείας· ἐκκλησία

Hesp 1933, pp. 156-158, no. 5   275/4 B.C.
(Meritt), lines 1-4:

Ἐπὶ Ὀλβίου ἄρχοντος ἐπὶ τῆς Λεωντίδος δωδεκάτ[ης]
[πρ]-
υτανείας ἧι Κυδίας Τιμωνίδου Εὐωνυμεὺς
ἐγρα[μμάτε]-
υεν· Σκιροφοριῶνος ἔνει καὶ νέαι, ἐνάτει καὶ
εἰ[κοστε͂]-
ι τῆς πρυτανείας· ἐκκλησία

Meritt, *Year*, pp. 192-194,   271/0 B.C.
lines 1-5:

['Ἐπὶ Πυθαράτου ἄρχοντος ἐπὶ τῆς Οἰ]νεῖδος [δωδε]-
[κάτης πρυτανείας ἧι Ἰσήγορος Ἰ]σοκράτου Κ[εφα]-
[λῆθεν ἐγραμμάτευεν· Σκιροφο]ριῶνος ἔνει καὶ [ν]-
[έαι προτέραι, μιᾶι καὶ τριακο]στῆι τῆς πρυτανε-
[ίας· ἐκκλησία]

IG II² 916 (see also Pritchett and   192/1 B.C.
Meritt, *Chronology*, pp. 115-116),
lines 8-11:

['Ἐπὶ ...ᶜᵃ· ¹⁰... ἄρχοντος τοῦ μετ]ὰ Φαναρχίδην ἐπὶ
τῆς
[Π ...ᶜᵃ· ⁸... ίδος δωδεκάτης πρυτ]ανείας ἧι Προκλῆς
Περι-
[......ᶜᵃ· ¹⁵...... ἐγραμμάτευεν·] Σκιροφοριῶνος ἔνει
καὶ νέ-
[αι, τριακοστῆι τῆς πρυτανείας· ἐκ]κλησία ἐν τῶι
θεάτρωι

III. THE ATHENIAN CALENDAR

Hesp 1934, pp. 18–21, no. 18   169/8 B.C.
(Meritt), lines 2–7:

Ἐπὶ Εὐνίκου ἄρχοντος ἐπὶ τῆς Ἀτταλίδος δωδεκάτης πρυτανείας ἧι Ἱερώνυ-
μος Βοήθου Κηφισιεὺς ἐγραμμάτευεν·
Σκιροφοριῶνος ἕνει καὶ νέαι, ἐνάτει
καὶ εἰκοστῆι τῆς πρυτανείας· ἐκκλησία ἐμ Πειραιεῖ

IG II² 945, lines 2–5:   168/7 B.C.

Ἐπὶ Ξενοκλέους ἄρχοντος ἐπὶ τῆς Οἰνεῖδος δωδεκάτης
πρυτα-
νείας ἧι Σθενέδημος Ἀσκ(λ)ηπιάδου Τειθράσιος
ἐγραμμάτευεν·
Σκιροφοριῶνος ἕνει καὶ νέαι, μιᾶι καὶ τριακοστῆι
τῆς πρυ[τανείας·]
ἐκκλησία σύγκλητος ἐν τῶι θεάτρωι

According to Lysias 26.6 a sacrifice was made to Zeus Soter on the last day of the year: ἡ γὰρ αὔριον ἡμέρα μόνη λοιπὴ τοῦ ἐνιαυτοῦ ἐστιν, ἐν δὲ ταύτῃ τῷ Διὶ τῷ σωτῆρι θυσία γίγνεται. This sacrifice should not be associated with the Diisoteria which was celebrated near the end of the year.[62] *IG* II² 1496, lines 117–121,

[ἐγ Βενδιδ]έων παρὰ ἱε[ροποιῶν: -]
[ἐκ τῆς θυσ]ίας τῶι Δι[ὶ τ]ῶι [Σωτῆρι]
[παρὰ βοων]ῶν: XXΓ̄ΗΔΙΙΙ
[ἐκ ..⁵.. ω]ν παρὰ βοώνου: Η[-]
[ἐκ ....⁸...]ων παρ[ὰ ἱ]εροπο[ιῶν]

indicates that at least two other public festivals came between the Diisoteria and the end of the year, and therefore the Diisoteria could not have been on Skirophorion 30.

The ten meetings of the Ekklesia reveal that Skirophorion 30 was a very common meeting day, and one may infer that a meeting of the Ekklesia occurred on Skirophorion 30 every year. The offering to which Lysias refers was probably made shortly before the meeting began, perhaps as a special addition to the sacrifices that regularly preceded meetings of the Ekklesia.

[62] Deubner, *Feste*, pp. 174–176.

SKIROPHORION

For a financial transaction on this day, see Skirophorion 30 in Appendix I.

Restorations to give Skirophorion 30:

A meeting of the Ekklesia in 341/0 B.C., *IG* II^2 229 + addenda, lines 2-5. See also Meritt, *Year*, p. 10.

A meeting of the Ekklesia in 337/6 B.C., *IG* II^2 242, lines 1-5.

A meeting of the Ekklesia in 337/6 B.C., *IG* II^2 276, lines 1-5 as restored by Schweigert, *Hesp* 1940, p. 342.

A meeting of the Ekklesia in 336/5 B.C., *IG* II^2 330, lines 47-49.

A meeting of the Ekklesia in 335/4 B.C., *IG* II^2 331, lines 1-5 as restored by Meritt, *Year*, p. 81. These lines are restored in *IG* II^2 331 to give Boedromion 30.

A meeting of the Ekklesia in 322/1 B.C., *IG* II^2 375, lines 1-7 as corrected by all editors from Thargelion 30. See Thargelion 30.

A meeting of the Ekklesia in 319/8 B.C., *Hesp* 1941, pp. 268-270, no. 69 (Pritchett), lines 1-5.

A meeting of the Ekklesia in 304/3 B.C., *IG* II^2 597 addenda, lines 2-3.

A meeting of the Boule in 303/2 B.C., *Hesp* 1952, pp. 367-368, no. 8 (Meritt), lines 1-7.

A meeting of the Ekklesia in 283/2 B.C., *Hesp* 1940, pp. 84-85, no. 15 (Meritt), lines 2-6.

A meeting of the Ekklesia in 169/8 B.C., *IG* II^2 911, lines 1-6.

Skirophorion—Summary

The numerous meetings which occurred during this month, and especially on the last day of the month, suggest that the Athenian government like all governments allowed business to accumulate during its term and then attempted to settle all of it hastily near the end of its term.

From the ten meetings attested for Skirophorion 30, more than twice as many as attested for any other day of the year, we may infer that there was a meeting of the Ekklesia on Skirophorion 30 every year. This is also indicated by the fact that no meeting is attested for Skirophorion 29. The Ekklesia would rarely meet two days in succession, and if it regularly met on Skirophorion 30, one would not expect to find meetings dated to Skirophorion 29.

CHAPTER IV
CONCLUSIONS

1. MEETING DAYS OF THE EKKLESIA

Distribution in the Year

Aristotle (*Ath. Pol.* 43.3) states that the Ekklesia met four times during each prytany, and we therefore might expect that the preserved decrees of the Ekklesia would be distributed fairly evenly among the twelve months. A survey of the total number of positively dated meetings of the Ekklesia in each month indicates, however, that they are very unevenly distributed.

Hek	Met	Boe	Pya	Mai	Pos	Gam	Anth	Ela	Mou	Tha	Ski
5	12	13	10	4	8	10	7	25	11	14	26

The most consistent disparity is between what might be termed the "winter months" Maimakterion through Anthesterion (October/November through January/February) and the "summer months" Elaphebolion through Pyanopsion (February/March through September/October). The winter months make up one-third of the year, and yet contain only one-fifth of the meetings.

One explanation of this uneven distribution would be that more meetings of the Ekklesia regularly occurred in some months than in others, but this would contradict Aristotle's explicit statement that four meetings of the Ekklesia occurred during each prytany. An alternative explanation of this uneven distribution may be suggested. The majority of the preserved decrees are honorary, and both citizens and foreigners would have preferred to receive honors during the summer months. During the winter months travel, and especially sea travel, was difficult, and this would limit the attendance at the Ekklesia and the recognition of the honor. A foreigner would have wished to be in Athens to receive his honor personally, and in many cases he could travel to Athens safely only during the summer months. For this reason, it may be suggested, we have proportionately more dated decrees from the summer months than from the winter months.

EKKLESIA MEETING DAYS

Distribution in the Month

Now that the positively dated meetings of the Ekklesia are isolated and are plotted on a calendar, it is possible to determine whether the four meetings specified by Aristotle could occur on any day of the month, or only on certain specified days, or whether the meeting days tended to follow a general pattern.

The distribution of positively dated meetings of the Ekklesia according to days of the month is as follows:

Day	Number of Meetings	Day	Number of Meetings
1	0	16	8
2	0	17	0
3	0	18	12
4	0	19	8
5	2	20	0
6	0	21	11
7	0	22	4
8	1	23	8
9	11	24	1
10	3	25	8
11	22	26	0
12	4	27	3
13	1	28	1
14	4	29	13
15	0	30	20

This distribution, viewed in conjunction with the calendar of attested meeting days of the Ekklesia (Calendar I) allows two immediate conclusions: first, the Ekklesia did not meet on every day of the month. There are ten days for which no meetings are attested —a number far too great to be solely the result of chance. Secondly, the Ekklesia did not invariably meet on a certain four specified days each month.

There are certain discernible patterns to the days of the meetings. The relatively small number of meetings in the first eight days is noteworthy. As will be seen, several of these first eight days were monthly festival days. Also, some time was required for the installation and organization of new prytaneis, who, in the period of twelve tribes, usually assumed office at the beginning of the month.

IV. CONCLUSIONS

CALENDAR I

Attested Meeting Days of the Ekklesia

	Hek	Met	Boe	Pya	Mai	Pos	Gam	Anth	Ela	Mou	Tha	Ski
1												
2												
3												
4												
5						E			E			
6												
7												
8									E			
9		E	E			E	E		E			E
10			E								E	E
11	E	E	E	E	E	E	E			E	E	E
12	E	E							E	E		
13									E			
14			E						E		E	
15												
16				E						E	E	E
17												
18			E	E				E	E		E	E
19						E		E	E	E		
20												
21		E			E				E			E
22				E					E	E		
23		E							E		E	E
24		E										
25			E	E		E	E		E			E
26												
27			E				E					E
28								E				
29		E		E		E	E		E	E	E	
30			E		E		E	E	E	E		E

E = attested meeting day of Ekklesia

These two factors contributed to limiting the number of meetings of the Ekklesia in the first eight days of the month.

The eleventh day of the month was clearly a regular meeting day. Twenty-two meetings of the Ekklesia are attested for this day, and they are distributed over ten of the twelve months (see Calendar I). Anthesterion 11 was a day of the Anthesteria, and Elaphebolion 11 was a day of the City Dionysia. These are the only two months for which no meetings are attested on the eleventh day, and thus they form the type of exception that tends to prove the rule.

The distribution of meetings also suggests that one of the last two days of a month was regularly a meeting day of the Ekklesia. The thirty-three meetings on the last two days are distributed over eleven of the twelve months. Only for Hekatombaion is there no meeting attested for one of the last two days. The Panathenaia, of which the chief day was Hekatombaion 28, may have precluded the possibility of a meeting on Hekatombaion 29 and 30.

The eleventh day and either the twenty-ninth or thirtieth day of a month may therefore be designated as regular meeting days of the Ekklesia, and, in general terms, they account for two of the four meetings specified by Aristotle. No clear pattern emerges for the two other meetings. It might be suggested that one meeting regularly occurred during days 16–19, and another during days 21–25, but the evidence is not conclusive.

The preferred days for meetings do not seem to have changed when the number of tribes, and thus of prytanies, was enlarged from ten to twelve. Of the twenty-six attested meeting days of the Ekklesia during the period of ten tribes, five (19 percent) were the eleventh day of a month and five (19 percent) were the twenty-ninth or thirtieth day. Of the sixty-six attested meeting days of the Ekklesia during the periods of twelve tribes, seven (11 percent) were the eleventh day of a month and thirteen (20 percent) were one of the last two days of a month. Likewise, of the thirty attested meetings of the Ekklesia during the period of ten tribes, seven (23 percent) occurred on the eleventh day and six (20 percent) on the twenty-ninth or thirtieth day. Of the 114 attested meetings of the Ekklesia in the periods of twelve tribes, fourteen (12 percent) occurred on the eleventh day, and twenty-seven (24 percent) on one of the last two days of a month. Although the sample for each period is admittedly small and inadequate, these statistics do suggest that the preferred

IV. CONCLUSIONS

days for meetings did not change radically from the period of ten tribes to the periods of twelve tribes.

2. Festival Days and Meeting Days of the Ekklesia

The fundamental question whether meetings of the Ekklesia frequently occurred on festival days may now be examined in view of the evidence from the calendar of positively dated festivals and meetings. The positively dated festival days and meeting days of the Ekklesia are tabulated in Calendar II.

The following 111 festival days for which no meeting of the Ekklesia is attested substantiate the hitherto tentative conclusion (see p. 7) that meetings of the Ekklesia did not frequently occur on festival days.

Festival	*Month and Day(s)*	*Total Days*
Noumenia	first day of every month	12
Day of Agathos Daimon (Niketeria)	second day of every month (Boedromion 2)[1]	12
Athena's day (Plataea celebration)	third day of every month (Boedromion 3)	12
Day of Herakles, Hermes, Aphrodite, and Eros	fourth day of every month	12
Artemis' day (Artemis Agrotera) (Delphinia)	sixth day of every month (Boedromion 6) (Mounichion 6)	12
Apollo's day (Pyanopsia) (Thargelia)	seventh day of every month (Pyanopsion 7) (Thargelion 7)	12
Day of Poseidon and Theseus (Theseia)	eighth day of every month[2] (Pyanopsion 8)	11

[1] When a monthly festival day and an annual festival day fell on the same day, they naturally formed only one festival day of the year. Therefore the annual festivals which coincide with monthly festivals are listed under the appropriate monthly festivals.

[2] For Elaphebolion 8 see *infra*.

FESTIVALS/ EKKLESIA MEETING DAYS

Festival	Month and Day(s)	Total Days
Synoikia	Hekatombaion 15 and 16	2
Panathenaia	Hekatombaion 28	1
Genesia	Boedromion 5	1
Democratia	Boedromion 12	1
Eleusinian Mysteries	Boedromion 15–17, 19–22	7
Stenia	Pyanopsion 9	1
Thesmophoria in Halimus	Pyanopsion 10	1
Nesteia	Pyanopsion 12	1
Kalligeneia	Pyanopsion 13	1
Chalkeia	Pyanopsion 30	1
Haloa	Posideon 26	1
Anthesteria	Anthesterion 11–13	3
Diasia	Anthesterion 23	1
City Dionysia	Elaphebolion 10–11	2
Bendideia	Thargelion 19	1
Plynteria	Thargelion 25	1
Skira	Skirophorion 12	1
Dipolieia	Skirophorion 14	1

On the following festival days, however, meetings of the Ekklesia are attested to have occurred.

Festival	Month and Day(s)	Number of Meetings
Kronia	Hekatombaion 12	1
Anodos	Pyanopsion 11	1
Theogamia	Gamelion 27	1
Proagon and sacrifice to Asklepios	Elaphebolion 8	1
City Dionysia	Elaphebolion 12–14	3
Mounichia	Mounichion 16	1
Olympieia	Mounichion 19	2

The total number of positively dated festival days (i.e., the total in the two lists) is 120, which constitutes 33 percent of the days of the year; the total number of meetings which conflict with festival

IV. CONCLUSIONS

CALENDAR II

Meeting Days of the Ekklesia and Festival Days

	Hek	Met	Boe	Pya	Mai	Pos	Gam	Anth	Ela	Mou	Tha	Ski
1	F	F	F	F	F	F	F	F	F	F	F	F
2	F	F	F	F	F	F	F	F	F	F	F	F
3	F	F	F	F	F	F	F	F	F	F	F	F
4	F	F	F	F	F	F	F	F	F	F	F	F
5			F			E			E			
6	F	F	F	F	F	F	F	F	F	F	F	F
7	F	F	F	F	F	F	F	F	F	F	F	F
8	F	F	F	F	F	F	F	F	E F	F	F	F
9		E	E	F		E	E		E			E
10			E	F					F		E	E
11	E	E	E	E F	E	E	E	F	F	E	E	E
12	E F	E	F	F				F	E F	E		F
13				F				F	E F			
14			E						E F		E	F
15	F		F									
16	F		F	E						E F	E	E
17			F									
18			E	E				E	E		E	E
19			F			E		E	E	E F	F	
20			F									
21		E	F		E				E			E
22			F	E					E	E		
23		E						F	E		E	E
24		E										
25			E	E		E	E		E		F	E
26						F						
27			E				E F					E
28	F							E				
29		E		E		E	E		E	E	E	
30			E	F	E		E	E	E	E		E

E = attested meeting day of Ekklesia
F = attested festival day

days is ten, which constitutes ca. 7 percent of the positively dated meetings. This alone indicates how infrequently meetings of the Ekklesia occurred on festival days. But this bare statistical summary is insufficient: other relevant factors must be considered concerning the ten meetings which are attested for festival days.

The meetings of the Ekklesia on Hekatombaion 12 (Kronia) and Elaphebolion 8 (Proagon and sacrifice to Asklepios) are recorded by the sources (Demosthenes 24.26 and Aeschines 3.66-67) as unusual and exceptional, precisely because meetings on these festival days were irregular (see Hekatombaion 12 and Elaphebolion 8 in the calendar). These two instances of conflict must, therefore, be viewed somewhat differently from the other instances. The conclusion to be drawn from these two instances is that meetings on festival days were considered irregular.

Meetings of the Ekklesia were, of course, attended solely by men, and may have been held during some of the festivals celebrated exclusively by women. The exclusion of men from the Thesmophoria is emphasized by Aristophanes throughout the *Thesmophoriazusai*, and a meeting of the Ekklesia in the theater is attested for the first day of this festival.[3] On this day, the Anodos, the women went in procession to the Thesmophorion on the Pnyx.[4] Since the women met at the Pnyx, the meeting of the Ekklesia necessarily was held in the theater. The situation of men holding a meeting while the women celebrated the Thesmophoria was not unique to Athens. Xenophon, *Hell.* 5.2.29, describes how the Theban men held a council while the women celebrated the Thesmophoria. Here too the Theban women displaced the men from their regular meeting place. It should also be noted that no meetings are attested for Pyanopsion 12, the most sacred day of the Thesmophoria.

The Theogamia celebrated on Gamelion 27 was a festival of Hera in her role as protectress of marriage, and, although there is no specific evidence concerning participation in the festival, we may now suspect that it was celebrated exclusively by women, so that a meeting of the Ekklesia[5] could occur on this day.

The three meetings of the Ekklesia during the course of the City Dionysia suggest that meetings occurred not infrequently during this

---

[3] *IG* II² 1006, lines 50-51.
[4] Deubner, *Feste*, p. 54.
[5] *IG* II² 849, lines 1-4.

IV. CONCLUSIONS

festival. The most sacred day was Elaphebolion 10, the day of the procession.[6] For this day and the following day no meetings are attested. By the fourth century B.C. many secular elements had been introduced into the City Dionysia,[7] and the dramatic contests had lost much of their religious content. Possibly for this reason they were interrupted for meetings.

Three instances of conflict remain. Similar reasons might be suggested for these, but none so convincing as the reasons for the other conflicts. Some of these conflicts may result from scribal error in inscribing the date of a decree. Such errors are common,[8] but there is no clear evidence of error in these cases.

As has been shown, two of the ten instances of conflict are self-proclaimed exceptions, but even if we allow for all ten instances of conflict listed, 93 percent of the positively dated meetings of the Ekklesia occurred on non-festival days. We have seen that several of the exceptions allow a plausible explanation. And thus a final conclusion can be reached: with very few exceptions meetings of the Ekklesia did not occur on festival days.

The conclusion that with few exceptions meetings of the Ekklesia did not occur on festival days is based on the positively dated meetings and festivals. On the basis of this conclusion, one must doubt the following restorations of fragmentary texts to give meetings of the Ekklesia on festival days.

A meeting of the Ekklesia on Skirophorion 2 (day of Agathos Daimon) in 339/8 B.C., *Hesp* 1938, pp. 291–292, no. 18 (Schweigert), lines 1–3.

A meeting of the Ekklesia on Gamelion 7 (Apollo's day) in 337/6 B.C., *IG* II^2 239, lines 3–7 as restored by Schweigert, *Hesp* 1940, p. 327.

A meeting of the Ekklesia on Hekatombaion 16 (Synoikia) in 332/1 B.C., *IG* II^2 420, lines 1–4 as restored by Meritt, *AJP* 85 (1964), pp. 304–306.

A meeting of the Ekklesia on Anthesterion 11 (Anthesteria) in 331/0 B.C., *IG* II^2 363, lines 1–6 as restored by Meritt, *Year*, pp. 88–89.

A meeting of the Ekklesia on Thargelion 1 (Noumenia) in

---

[6] See Elaphebolion 9 in the calendar.

[7] Pickard-Cambridge, *The Dramatic Festivals of Athens* as revised by Gould and Lewis, pp. 58–59.

[8] Pritchett and Neugebauer, *Calendars*, p. 38.

FESTIVALS/ EKKLESIA MEETING DAYS

330/29 B.C., *IG* II^2 351 addenda, lines 2–7 as restored by Meritt, *Year*, pp. 91–94.

A meeting of the Ekklesia on Pyanopsion 30 (Chalkeia) in 329/8 B.C., *IG* II^2 353, lines 1–7.

A meeting of the Ekklesia on Elaphebolion 8 (Proagon and sacrifice to Asklepios) in 326/5 B.C., *IG* II^2 359, lines 2–7.

A meeting of the Ekklesia on Gamelion 6 (Artemis' day) in 320/19 B.C., *Hesp* 1944, pp. 234–241, no. 6 (Meritt), lines 2–6 as restored by Dow, *HSCP* 67 (1963), pp. 67–75.

A meeting of the Ekklesia on Mounichion 8 (day of Poseidon and Theseus) in 320/19 B.C., *Hesp* 1944, pp. 234–241, no. 6 (Meritt), lines 2–6 as restored by Meritt, *Year*, pp. 119–120.

A meeting of the Ekklesia on Skirophorion 14 (Dipolieia) in 319/8 B.C., *IG* II^2 390, lines 1–4 as restored by Dinsmoor, *Archons*, pp. 21–22.

A meeting of the Ekklesia on Thargelion 2 (day of Agathos Daimon) in 307/6 B.C., *IG* II^2 455, lines 1–4 as restored by Meritt, *Year*, pp. 177–178. Also *IG* II^2 460, lines 1–5 as restored by Pritchett and Meritt, *Chronology*, pp. 17–18.

A meeting of the Ekklesia on Skirophorion 3 (Athena's day) in 307/6 B.C., *IG* II^2 460, lines 1–5 as restored by Meritt, *Year*, pp. 177–178.

A meeting of the Ekklesia on Skirophorion 4 (day of Herakles, Hermes, Aphrodite, and Eros) in 307/6 B.C., *IG* II^2 455, lines 1–4 as restored by Pritchett and Meritt, *Chronology*, p. 20.

A meeting of the Ekklesia on Anthesterion 8 (day of Poseidon and Theseus) in 303/2 B.C., *IG* II^2 489, lines 1–5.

A meeting of the Ekklesia on Skirophorion 6 (Artemis' day) in 303/2 B.C., *IG* II^2 498 + addenda, lines 1–6.

A meeting of the Ekklesia on Posideon 4 (day of Herakles, Hermes, Aphrodite, and Eros) in 294/3 B.C., *IG* II^2 378, lines 1–5 as restored by Pritchett and Neugebauer, *Calendars*, pp. 70–72.

A meeting of the Ekklesia on Elaphebolion 11 (City Dionysia) in 280/79 B.C., *IG* II^2 670, lines 1–4.

A meeting of the Ekklesia on Skirophorion 12 (Skira) in 280/79 B.C., *IG* II^2 670 addenda, lines 1–4.

A meeting of the Ekklesia on Gamelion 8 (day of Poseidon and Theseus) in 279/8 B.C., *Hesp* 1963, pp. 5–6, no. 6 (Meritt), lines 1–4 as restored by Meritt, *Hesp* 1969, p. 110.

A meeting of the Ekklesia on Elaphebolion 4 (day of Herakles,

IV. CONCLUSIONS

Hermes, Aphrodite, and Eros) in 279/8 B.C., *Hesp* 1948, pp. 1–2, no. 1 (Meritt), lines 1–4.

A meeting of the Ekklesia on Skirophorion 8 (day of Poseidon and Theseus) in 267/6 B.C., *IG* II^2 664 + addenda, lines 1–5.

A meeting of the Ekklesia on Posideon 2 (day of Agathos Daimon) in 263/2 B.C., *IG* II^2 477, lines 1–6 as restored by Meritt, *Hesp* 1938, pp. 141–142.

A meeting of the Ekklesia on Posideon 3 (Athena's day) in 255/4 B.C., *IG* II^2 447, lines 1–6 as restored by Meritt, *Hesp* 1969, pp. 435–436.

A meeting of the Ekklesia on Boedromion 20 (Eleusinian Mysteries) in 245/4 B.C., *IG* II^2 799, lines 2–4.

A meeting of the Ekklesia on Boedromion 12 (Demokratia) in 244/3 B.C., *Hesp* 1938, pp. 114–115, no. 21 (Meritt), lines 1–4.

A meeting of the Ekklesia on Boedromion 8 (day of Poseidon and Theseus) in 226/5 B.C., *AJP* 63 (1942), p. 422, lines 1–4 as restored by Meritt, *Year*, p. 154.

A meeting of the Ekklesia on Anthesterion 8 (day of Poseidon and Theseus) in 226/5 B.C., *AJP* 63 (1942), p. 422 (Pritchett), lines 1–4.

A meeting of the Ekklesia on Thargelion 19 (Bendideia) in 190/89 B.C., Pritchett and Meritt, *Chronology*, pp. 123–126, lines 1–4.

A meeting of the Ekklesia on Elaphebolion 10 (procession of the City Dionysia) in 189/8 B.C., Dow, *Prytaneis*, pp. 91–92, no. 41, lines 25–26 as restored by Meritt, *Hesp* 1957, pp. 63–64, no. 17.

A meeting of the Ekklesia on Pyanopsion 30 (Chalkeia) in 171/0 B.C., *Hesp* 1934, pp. 14–18, no. 17 (Meritt), lines 1–5.

A meeting of the Ekklesia on Mounichion 4 (day of Herakles, Hermes, Aphrodite, and Eros) in 128/7 B.C., *Hesp* 1935, pp. 71–81, no. 37 (Dow), lines 115–116 as restored by Meritt, *Hesp* 1946, pp. 201–213, no. 41.

A meeting of the Ekklesia on Boedromion 5 (Genesia) in 127/6 B.C., *Hesp* 1935, pp. 71–81, no. 37 (Dow), lines 1–3.

A meeting of the Ekklesia on Posideon 26 (Haloa) in 125/4 B.C., *IG* II^2 1003, lines 1–3 as restored by Meritt, *Year*, pp. 190–191.

A meeting of the Ekklesia on Maimakterion 6 (Artemis' day) in 118/7 B.C., *Hesp* 1963, pp. 22–23, no. 23 (Meritt), lines 1–4.

Alternative restorations for most of these texts have already been proposed by scholars (see specific days cited), and others might be suggested here. But although the restored versions listed *supra* can now with some certainty be regarded as incorrect, in no case can an alternative restoration be proved correct by the criteria specified on pp. 10–12.

3. MEETING DAYS OF THE BOULE

Aristotle (*Ath. Pol.* 43.3) states that the Boule met every day of the year, except if a day was ἀφέσιμος. The adjective ἀφέσιμος does not have the religious connotations of "holiday" or "festival" day, but rather blandly indicates a day of "recess" or "dismissal." The Boule, although it met much more frequently than the Ekklesia, seldom enacted resolutions which were to be inscribed on stone, and therefore only twenty-four meetings, including twenty-three different days, are positively attested. For the distribution of these see Calendar III.

The distribution of the attested meetings of the Boule in the year is as follows:

Hek	Met	Boe	Pya	Mai	Pos	Gam	Anth	Ela	Mou	Tha	Ski
1	3	3	2	4	1	3	1	0	1	2	3

The distribution of the attested meetings of the Boule in the month is as follows:

Day	Number of Meetings	Day	Number of Meetings
1	0	16	1
2	2	17	0
3	2	18	0
4	3	19	0
5	1	20	0
6	4	21	0
7	0	22	2
8	1	23	2
9	0	24	1
10	0	25	1
11	0	26	0
12	0	27	2
13	0	28	0
14	1	29	1
15	0	30	0

IV. CONCLUSIONS

CALENDAR III

Attested Meeting Days of the Boule

	Hek	Met	Boe	Pya	Mai	Pos	Gam	Anth	Ela	Mou	Tha	Ski
1												
2		B				B						
3										B		B
4		B	B									B
5					B							
6			B	B	B							
7												
8							B					
9												
10												
11												
12												
13												
14											B	
15												
16					B							
17												
18												
19												
20												
21												
22	B					B						
23											B	B
24			B									
25							B					
26												
27		B						B				
28												
29					B							
30												

B = attested meeting day of Boule

BOULE MEETING DAYS

Calendar IV

Attested Meeting Days of the Boule and of the Ekklesia

	Hek	Met	Boe	Pya	Mai	Pos	Gam	Anth	Ela	Mou	Tha	Ski
1												
2		B				B						
3										B		B
4		B	B									B
5					B	E			E			
6			B	B	B							
7												
8							B		E			
9		E	E			E	E		E			E
10			E								E	E
11	E	E	E	E	E	E	E			E	E	E
12	E	E							E	E		
13									E			
14			E						E		E B	
15												
16			E	B						E	E	E
17												
18			E	E				E	E		E	E
19					E			E	E	E		
20												
21		E		E					E			E
22	B		E			B			E	E		
23		E							E		E B	E B
24		E	B									
25			E	E		E	E B		E			E
26												
27		B	E			E	B					E
28								E				
29		E		E	B	E	E		E	E	E	
30			E		E		E	E	E	E		E

B = attested meeting day of the Boule
E = attested meeting day of the Ekklesia

IV. CONCLUSIONS

The number of attested meetings of the Boule is quite small, and one must take care in drawing conclusions from this limited sample. But the distribution of meetings of the Boule in the month contrasts sharply with the distribution of the meetings of the Ekklesia. This is seen most clearly in Calendar IV, which presents all the attested meeting days of the Boule and of the Ekklesia.

Fifty-four percent of the attested meetings of the Boule occur in the first eight days of the month, whereas only 2 percent of the attested meetings of the Ekklesia occur in the same period. This contrast is, obviously, significant. The reason for this contrast will become clear in the following discussion of the festival days and meeting days of the Boule. It is also noteworthy that for the eleventh and thirtieth days, the two most common meeting days of the Ekklesia, no meetings of the Boule are attested.

4. Festival Days and Meeting Days of the Boule

Aristotle, as discussed *supra*, stated that the Boule met every day, except if a day was ἀφέσιμος. Which days for the Boule were ἀφέσιμοι? As has been shown, the Ekklesia, with very few exceptions, did not meet on days of religious celebrations. Were these same days ἀφέσιμοι for the Boule? The answers to these questions lie in Calendar V, which lists all the attested festival days and all the attested meeting days of the Boule.

As is seen in Calendar V, twelve of the twenty-four (i.e., 50 percent) attested meetings of the Boule occurred on festival days. That is in striking contrast with the meetings of the Ekklesia, of which only 7 percent occurred on festival days. For an explanation of this surprising contrast, we must survey the individual instances of conflict between a meeting day of the Boule and a festival day.

Day	Festival	Number of Meetings of Boule
Metageitnion 2	day of Agathos Daimon	1
Metageitnion 4	day of Herakles, Hermes, Aphrodite, and Eros	1
Boedromion 4	day of Herakles, Hermes, Aphrodite, and Eros	1

MAJOR ATHENIAN FESTIVALS

Day	Festival	Number of Meetings of Boule
Boedromion 6	day of Artemis and festival of Artemis Agrotera	1
Pyanopsion 6	day of Artemis	2
Maimakterion 6	day of Artemis	1
Posideon 2	day of Agathos Daimon	1
Gamelion 8	day of Poseidon and Theseus	1
Mounichion 3	day of Athena	1
Skirophorion 3	day of Athena	1
Skirophorion 4	day of Herakles, Hermes, Aphrodite, and Eros	1

Eleven of the twelve meetings of the Boule which conflict with festival days occurred on monthly festival days as distinguished from annual festival days. Metageitnion 2 and 4, Boedromion 4, Pyanopsion 6, Maimakterion 6, Posideon 2, Gamelion 8, Mounichion 3, and Skirophorion 3 and 4 were all monthly festival days. Boedromion 6, besides being a monthly festival day devoted to Artemis, was also the annual festival day of Artemis Agrotera. If we distinguish between monthly festival days and annual festival days, the following statistics result: eleven of the twenty-four (ca. 42 percent) attested meetings of the Boule occurred on monthly festival days; only one of the twenty-four (ca. 4 percent) attested meetings of the Boule occurred on an annual festival day. And thus concerning the meetings of the Boule a conclusion can be reached. The Boule regularly met on monthly festival days, but, with very few exceptions, did not meet on annual festival days. The ἡμέραι ἀφέσιμοι for the Boule were, in all probability, only the annual festival days.

5. Probable Dates for Major Athenian Festivals

The conclusion that with very few exceptions meetings of the Ekklesia and meetings of the Boule did not occur on annual festival days has now been established. This conclusion provides a new approach to the dating of certain major Athenian festivals. With a certainty far greater than hitherto possible, one can now eliminate

IV. CONCLUSIONS

Calendar V

Meeting Days of the Boule and Festival Days

	Hek	Met	Boe	Pya	Mai	Pos	Gam	Anth	Ela	Mou	Tha	Ski
1	F	F	F	F	F	F	F	F	F	F	F	F
2	F	B F	F	F	F	B F	F	F	F	F	F	F
3	F	F	F	F	F	F	F	F	F	B F	F	B F
4	F	B F	B F	F	F	F	F	F	F	F	F	B F
5			F		B							
6	F	F	B F	B F	B F	F	F	F	F	F	F	F
7	F	F	F	F	F	F	F	F	F	F	F	F
8	F	F	F	F	F	F	B F	F	F	F	F	F
9			F									
10			F					F				
11			F					F	F			
12	F		F	F				F	F			F
13				F				F	F			
14									F		B	F
15	F		F									
16	F		F		B					F		
17			F									
18												
19			F							F	F	
20			F									
21			F									
22	B		F				B					
23								F			B	B
24			B									
25							B				F	
26						F						
27		B					F	B				
28	F											
29					B							
30				F								

F = attested festival day
B = attested meeting day of Boule

MAJOR ATHENIAN FESTIVALS

as probable annual festival days those days for which meetings of the Ekklesia or Boule are positively attested. These meeting days, together with the established festival days, are tabulated in Calendar VI.

On the basis of these same conclusions one can now also designate as possible festival days those days for which no meeting is attested. If the number of positively dated meetings which survive were ten or twenty times as great, single days for which no meeting is attested could be designated with virtual certainty as festival days. Because of the limited amount of evidence, however, the present discussion must be confined to sequences of three or more days.

Some major Athenian festivals require a continuous sequence of three or more festival days, i.e., three or more days in succession for which no meeting is attested. Such sequences are infrequent, and can often be associated with known festivals. For example, if a given four-day festival is attested to have occurred in a certain month, and if there is only one continuous sequence of four possible festival days in that month, then the festival should be dated to those days.

This method for dating major festivals has been used together with other evidence in the summaries of the months. The tentative conclusion upon which those discussions were based (see p. 7) has proved to be valid, and the following new or improved datings of festivals have resulted:

Panathenaia (For discussion see Hekatombaion 28 and Summary)

The central day was Hekatombaion 28. The festival also included the two following days, Hekatombaion 29 and 30. Although the length of the festival may have varied from year to year, it usually included Hekatombaion 23–30.

Eleusinia (For discussion see Metageitnion—Summary)

This festival included at least four days within the period Metageitnion 13–20. Since the celebrants traveled to and from Eleusis, the central days Metageitnion 15–18 were probably the festival days.

IV. CONCLUSIONS

CALENDAR VI

Meeting Days of Ekklesia and Boule and Festival Days

	Hek	Met	Boe	Pya	Mai	Pos	Gam	Anth	Ela	Mou	Tha	Ski
1	F	F	F	F	F	F	F	F	F	F	F	F
2	F	B F	F	F	F	B F	F	F	F	F	F	F
3	F	F	F	F	F	F	F	F	F	B F	F	B F
4	F	B F	B F	F	F	F	F	F	F	F	F	B F
5			F		B	E			E			
6	F	F	B F	B F	B F	F	F	F	F	F	F	F
7	F	F	F	F	F	F	F	F	F	F	F	F
8	F	F	F	F	F	F	B F	F	E F	F	F	F
9		E	E	F		E	E		E			E
10			E	F					F		E	E
11	E	E	E	E F	E	E	E	F	F	E	E	E
12	E F	E	F	F				F	E F	E		F
13				F				F	E F			
14			E						E F		B E	F
15	F		F									
16	F		F	E	B					E F	E	E
17			F									
18			E	E					E	E	E	E
19			F			E			E	E	E F	F
20			F									
21		E	F		E				E			E
22	B		F	E			B		E	E		
23		E						F	E		B E	B E
24		E	B									
25			E	E		E	B E		E		F	E
26						F						
27		B	E				E F	B				E
28	F							E				
29		E		E	B	E			E	E	E	
30			E	F	E		E		E	E		E

E = attested meeting day of Ekklesia
B = attested meeting day of Boule
F = attested festival day

MAJOR ATHENIAN FESTIVALS

Apatouria   (For discussion see Pyanopsion—Summary)

The Apatouria required a sequence of three festival days in Pyanopsion. There are two possibilities: Pyanopsion 19–21 and 26–28. No evidence available at present determines the choice.

Lenaia   (For discussion see Gamelion 19 and Summary)

This dramatic festival began on Gamelion 12 and continued, perhaps, until Gamelion 21. Apparently it was still in progress on Gamelion 19.

Mysteries at Agrai   (For discussion see Anthesterion—Summary)

The mysteries must have included at least three successive days in the period Anthesterion 20–26. Anthesterion 23 was one of the festival days.

City Dionysia   (For discussion see Elaphebolion 9–18 and Summary)

The festival began on Elaphebolion 10. Its duration may have varied, but in 346 B.C. the City Dionysia and Pandia together lasted until Elaphebolion 17.

In Calendar VII are tabulated all the positively attested festival days and meeting days of the Ekklesia and Boule. In addition the new or improved datings of festivals are included in parentheses.

The total number of festival days appears surprisingly great. One hundred and twenty monthly and annual festival days are positively established, and to this number should be added the twenty-four festival days for which new dates have been proposed. To this total must also be added those festival days which are not firmly dated, such as the Apatouria, the Mysteries at Agrai, and other such. The total, if all the festival days were known, would probably be only slightly less than one-half of all the days of the year. But even this would not surpass the number of festival days of the Tarentians, who, according to Strabo (6.280), at the height of their prosperity had more public festival days in the year than non-festival days. Athens,

IV. CONCLUSIONS

Calendar VII

Meeting Days of the Ekklesia and Boule, Attested and Probable Festival Days

	Hek	Met	Boe	Pya	Mai	Pos	Gam	Anth	Ela	Mou	Tha	Ski
1	F	F	F	F	F	F	F	F	F	F	F	F
2	F	B F	F	F	F	B F	F	F	F	F	F	F
3	F	F	F	F	F	F	F	F	F	B F	F	B F
4	F	B F	B F	F	F	F	F	F	F	F	F	B F
5			F		B	E			E			
6	F	F	B F	B F	B F	F	F	F	F	F	F	F
7	F	F	F	F	F	F	F	F	F	F	F	F
8	F	F	F	F	F	F	B F	F	E F	F	F	F
9		E	E	F		E	E		E			E
10			E	F					F		E	E
11	E	E	E	E F	E	E	E	F	F	E	E	E
12	E F	E	F	F			(F)	F	E F	E		F
13				F			(F)	F	E F			
14		E					(F)		E F		B E	F
15	F	(F)	F				(F)		(F)			
16	F	(F)	F	E	B		(F)		(F)	E F	E	E
17		(F)	F				(F)		(F)			
18		(F)	E	E			(F)	E	E		E	E
19			F			E	(F)	E	E	E F	F	
20			F				(F)					
21		E	F		E		(F)		E			E
22	B		F	E			B		E	E		
23	(F)	E						F	E		B E	B E
24	(F)	E	B									
25	(F)		E	E		E	B E		E		F	E
26	(F)					F						
27	(F)	B	E				E F	B				E
28	F							E				
29	(F)	E		E	B	E	E		E	E	E	
30	(F)		E	F	E		E	E	E	E		E

E = attested meeting day of Ekklesia
B = attested meeting day of Boule
F = attested festival day   (F) = probable festival day

it must be remembered, prided herself on her numerous festival days, and was internationally renowned for them.

One must take care, however, to establish the proper relationship between "festival days" and "working days." They are by no means mutually exclusive in all cases. Both monthly and annual festival days have been proved to be non-working days vis-à-vis the Ekklesia. But for the Bouleutai only the annual festival days were non-working days. There is abundant evidence detailed *supra* that the Bouleutai regularly met on monthly festival days. Aristophanes (*Vesp*. 660–663), who is surely exaggerating, indicates that the Dikasts worked three hundred days each year. In general it would seem reasonable that those of the lower economic strata would be less able to enjoy festival days as non-working days. This assumption is strikingly confirmed by the record of the work crew at Eleusis (*IG* II^2 1672, lines 32–33, cited for Hekatombaion 4). These men began work on Hekatombaion 4 and were paid, and assuredly thus worked, for forty successive days. They not only labored on the monthly festival days in Hekatombaion and Metageitnion, but they even labored through the Kronia, the Synoikia, and the splendid festival of the Panathenaia. Clearly all festival days were not days of vacation for all the populace. And thus in reference to the results of this entire study an important point must be made: this study has concentrated solely on the relationship of festival days vis-à-vis meetings of the Boule and Ekklesia. Every day established as a festival day was not a day of vacation for all Athenians.

With very few exceptions meetings of the Ekklesia did not occur on monthly or annual festival days. Likewise meetings of the Boule did not occur on annual festival days. These newly discovered facts will require students of Greek epigraphy and Greek religion to treat the Athenian calendar as a unity of interrelated sacred and civil elements. Sacred and civil were so tightly interwoven in ancient Athenian life that consideration of one without the other has led to error. The epigraphists must now use the calendar of religious festival days as a new criterion for proposing and evaluating restorations. The historian of Greek religion can now use the calendar of meetings to isolate possible festival days.

The calendric study of meetings and festivals presented *supra* is a prerequisite for these studies. In the future this calendar will be augmented and improved by the discovery of new epigraphical texts

IV. CONCLUSIONS

and by the acceptance of restored texts which can be *proved* to be correct. The calendar as summarized in Calendar VI contains only positive evidence, and, if this calendar is preserved from the bottomless sea of hypothetical and unproved restorations, it will be a valuable source of new data for the study of Greek epigraphy and Greek religion.

APPENDICES

APPENDIX I

CALENDAR OF DATED FINANCIAL TRANSACTIONS

NUMEROUS financial transactions are recorded on Attic inscriptions, and some few of these are dated to the day and month on which they occurred. The largest group of precisely dated financial transactions which survive is from the accounts of the treasurers of Athena (*IG* I^2 304 B). These accounts detail the money which the treasurers of the treasury of Athena loaned to the Athenian state.

The dated financial transactions have been relegated to an appendix because they were handled by only a few officials, and do not presuppose a meeting of a legislative body. They therefore must be distinguished from the meetings of legislative assemblies presented in the calendar *supra*. The financial transactions have been included in this appendix primarily for the purpose of reference, but also in the hope that more inscriptions of the type of *IG* I^2 304 B may be uncovered, and a fruitful study of the relation of financial transactions to the civil and sacred calendar may then be undertaken.

The inscription which provides most of the positively dated financial transactions, *IG* I^2 304 B, has recently been the subject of intensive study. Pritchett, after devoting weeks of study to the stele itself, has published (*The Choiseul Marble*, University of California Publications: Classical Studies, Vol. 5, 1970) a new text, and a lengthy discussion of the text and of Greek calendric studies in general. By utilizing all the methods of modern epigraphy, Pritchett was able to read far more letters on the stele than either Hiller (in his text in *IG*) or Meritt (*AFD*, pp. 119–122). Although such radically new texts usually occasion debate, Meritt (*TAPA* 95, 1964, pp. 204–205) accepted with few exceptions Pritchett's new text. For this reason Pritchett's text (*Choiseul*, pp. 7–10) has been presented in the calendar *infra*, with the occasional variants of Meritt noted. Although Pritchett renumbers the lines, I have presented for the reader's convenience the line numbers as they occur in *IG*.

APPENDIX I

A problem in *IG* I² 304 B which is momentous for the interpretation of the text and for various calendric problems, but which is of no immediate importance for the purposes of this study, is the proper year date of the distinct parts of the text. Meritt (*AFD*, pp. 116–118, and *TAPA* 95, 1964, pp. 204–205) dates lines 41–92 to 407/6 B.C. Pritchett (*Choiseul*, pp. 22–25) dates lines 41–65 to 407/6 B.C. and lines 66–92 to 408/7 B.C.[9] Pritchett's arguments for his dating appear more cogent, and thus I have accepted his dating. This decision does not, however, affect the evidence for the purposes of this study.

Hekatombaion 2

Kirchner restored *IG* II² 1589, lines 1–4 to give a financial transaction of the poletai on this day in 307/6 B.C. The month, however, equally well may be restored to give Metageitnion 2.

Hekatombaion 14

IG II² 1492, lines 124–126 records a financial transaction of the treasurers of Athena and the other gods on this day in 305/4 B.C.

Hekatombaion 20

IG I² 304 B, lines 88–91 as read by Pritchett (*Choiseul*, p. 9) records two financial transactions of the treasurers of Athena on this day in 407/6 B.C. Meritt (*AFD*, p. 122) does not attempt to restore the day date in line 89, but gives the same day date in line 91.

Metageitnion 8 and 20

IG I² 304 B, lines 91–92 and 41–43 as read by Pritchett (*Choiseul*, p. 9) records single financial transactions of the treasurers of Athena on each of these days in 407/6 B.C.

---

[9] The bibliography for this controversy has been extensive. I have cited only the most important and/or the most recent studies, studies which summarize previous discussions. Pritchett's monograph (*The Choiseul Marble*), for example, synthesizes or slightly revises his previous work on the subject (such as *BCH* 88, 1964, pp. 455–481). Ferguson's contribution to this discussion (*The Treasurers of Athena*, Cambridge, Mass., 1932, pp. 28–32) should be noted.

FINANCIAL TRANSACTIONS

Metageitnion 25

IG I^2 304 B, lines 43–47 as read by Pritchett (*Choiseul*, p. 9) records two financial transactions of the treasurers of Athena on this day in 407/6 B.C.

Metageitnion 26, 27, and 30

IG I^2 304 B, lines 47–52 as read by Pritchett (*Choiseul*, p. 9) records single financial transactions of the treasurers of Athena on each of these days in 407/6 B.C.

Boedromion 1, 2, and 4

IG I^2 304 B, lines 52–57 as read by Pritchett (*Choiseul*, pp. 9–10) records single financial transactions of the treasurers of Athena on each of these days in 407/6 B.C.

Boedromion 8 and 14

IG I^2 304 B, lines 57–65 as read by Pritchett (*Choiseul*, p. 10) records two financial transactions of the treasurers of Athena on each of these days in 407/6 B.C.

Gamelion 7, 22, and 25

IG I^2 328, lines 5–18 records private sales on each of these days in 414/3 B.C.

Mounichion 3

IG I^2 304 B, lines 68–69 as read by Pritchett (*Choiseul*, p. 8) records a financial transaction of the treasurers of Athena on this day in 408/7 B.C.

Mounichion 4

IG II^2 1682, lines 1–2 records a financial transaction at Eleusis on this day in 285/4 B.C.

APPENDIX I

Mounichion 6

IG I² 304 B, line 70 as read by Pritchett (*Choiseul*, p. 8) records a financial transaction of the treasurers of Athena on this day in 408/7 B.C.

Mounichion 10

Hesp 1941, pp. 14–27, no. 1 (Crosby), lines 6–8 records a confiscation of property, a form of financial transaction, on this day in 367/6 B.C.

Mounichion 17

IG I² 304 B, lines 71–72 as read by Pritchett (*Choiseul*, p. 8) records a financial transaction of the treasurers of Athena on this day in 408/7 B.C.

Mounichion 18

IG I² 304 B, lines 72–74 would appear to record a financial transaction of the treasurers of Athena on Mounichion 8 in 408/7 B.C. But Pritchett (*Choiseul*, p. 12) notes that the sequence of days in the inscription indicates that the scribe must have erroneously omitted ἐπὶ δέκα after ὀγδόει in line 73. The proper date must be Mounichion 18. Meritt (*AFD*, p. 121) restores these lines to give Mounichion 19.

Mounichion 22

Kirchner interprets *IG* II² 1492, lines 97–99 so as to give a financial transaction of the treasurers of Athena and the other gods on this day in 306/5 B.C.

Mounichion 25

IG I² 304 B, lines 74–75 as read by Pritchett (*Choiseul*, p. 8) records a financial transaction of the treasurers of Athena on this day in 408/7 B.C. Meritt (*AFD*, p. 121) restores these lines to give Mounichion 22.

FINANCIAL TRANSACTIONS

Mounichion 28

IG I² 305, line 14 records a financial transaction of the treasurers of Athena on this day in 406/5 B.C.

Thargelion 2

IG I² 304 B, lines 76–77 as read by Pritchett (*Choiseul*, p. 8) records a financial transaction of the treasurers of Athena on this day in 408/7 B.C. Meritt originally (*AFD*, p. 121) restored line 77 to give Mounichion 30, but later (*TAPA* 95, 1964, p. 206) accepted Pritchett's reading to give Thargelion 2.

Thargelion 11

IG I² 304 B, line 79 as read by Pritchett (*Choiseul*, p. 8) records a financial transaction of the treasurers of Athena on this day in 408/7 B.C. Meritt originally (*AFD*, p. 121) restored this line to give Thargelion 7, but later (*TAPA* 95, 1964, p. 206) accepted Pritchett's reading to give Thargelion 11.

Thargelion 24

Kirchner has interpreted *IG* II² 1492, lines 104–105 to give a financial transaction of the treasurers of Athena and the other gods on this day in 306/5 B.C. The calculation of the date, however, depends upon a prytany number which has been restored.

Skirophorion 5 and 16

IG I² 304 B, lines 79–86 as read by Pritchett (*Choiseul*, p. 8) records two financial transactions of the treasurers of Athena on Skirophorion 5 and one transaction on Skirophorion 16 in 408/7 B.C.

Skirophorion 22

IG II² 1492, lines 118–120 records a financial transaction of the treasurers of Athena and the other gods on this day in 306/5 B.C. The number of the prytany establishes that the month was Skirophorion.

APPENDIX I

Skirophorion 24

IG I² 304 B, lines 83–85 as read by Pritchett (*Choiseul*, p. 8) records a financial transaction of the treasurers of Athena on this day in 408/7 B.C. *IG* II² 1492, lines 112–114 records a financial transaction on this day in 306/5 B.C.

Skirophorion 30

IG I² 304 B, lines 87–88 as read by Pritchett (*Choiseul*, pp. 8–9) records a financial transaction of the treasurers of Athena on this day in 408/7 B.C. Meritt (*AFD*, p. 121) restores these lines to give Skirophorion 26.

SUMMARY

Thirty-eight financial transactions are positively dated. In Calendar VIII the days of these financial transactions are tabulated together with the meeting days and festival days.

Eleven of the thirty-eight (ca. 29 percent) of the attested financial transactions occurred on festival days. The following are the ten festival days for which financial transactions are attested:

Metageitnion 8	day devoted to Poseidon and Theseus
Boedromion 1	the Noumenia
Boedromion 2	Niketeria and day of the Agathos Daimon
Boedromion 4	day of Herakles, Hermes, Aphrodite, and Eros
Boedromion 8	day of Poseidon and Theseus
Gamelion 7	Apollo's day
Mounichion 3	Athena's day
Mounichion 4	day of Herakles, Hermes, Aphrodite, and Eros
Mounichion 6	procession to Delphinion
Thargelion 2	day devoted to Agathos Daimon

Nine of the eleven financial transactions attested for festival days occurred on monthly festival days. Two of the eleven occurred on annual festival days (Boedromion 2 and Mounichion 6). Thus

FINANCIAL TRANSACTIONS

Calendar VIII

Meeting Days, Festival Days, and Days of Financial Transactions

	Hek	Met	Boe	Pya	Mai	Pos	Gam	Anth	Ela	Mou	Tha	Ski
1	F	F	F T	F	F	F	F	F	F	F	F	F
2	F	B F	F T	F	F	B F	F	F	F	F	F T	F
3	F	F	F	F	F	F	F	F	F	BFT	F	B F
4	F	B F	BFT	F	F	F	F	F	F	F T	F	B F
5		F		B	E				E			T
6	F	F	B F	B F	B F	F	F	F	F	F T	F	F
7	F	F	F	F	F	F	F T	F	F	F	F	F
8	F	F T	F T	F	F	F	B F	F	E F	F	F	F
9		E	E	F		E	E		E			E
10			E	F					F	T	E	E
11	E	E	E	E F	E	E	E	F	F	E	E T	E
12	E F	E	F	F			(F)	F	E F	E		F
13				F			(F)	F	E F			
14	T		E T				(F)		E F		B E	F
15	F	(F)	F				(F)		(F)			
16	F	(F)	F	E	B		(F)		(F)	E F	E	E T
17		(F)	F				(F)		(F)	T		
18		(F)	E	E			(F)	E	E	T	E	E
19			F			E	(F)	E	E	E F	F	
20	T	T	F				(F)					
21		E	F		E		(F)		E			E
22	B		F	E			B T		E	E		T
23	(F)	E						F	E		B E	B E
24	(F)	E	B									T
25	(F)	T	E	E		E	BET		E	T	F	E
26	(F)	T				F						
27	(F)	B T	E				E F	B				E
28	F							E		T		
29	(F)	E		E	B	E	E		E	E	E	
30	(F)	T	E	F	E		E	E	E	E		E T

F = attested festival day (F) = probable festival day
E = attested meeting day of Ekklesia
T = attested day of financial transaction
B = attested meeting day of Boule

APPENDIX I

24 percent of the attested financial transactions occurred on monthly festival days, whereas only 5 percent occurred on annual festival days. The conclusion concerning the financial transactions, just as concerning the meetings of the Boule, is that they quite regularly occurred on monthly festival days, but very infrequently on annual festival days.

APPENDIX II
PSEUDO-PSEPHISMATA

IT WAS THE standard practice in Athenian courts of law for the speaker to request the clerk of the court to read documents in support of the speaker's argument. These documents included nomoi, psephismata, affidavits of witnesses, and other documents of public record.

In the corpus of the speeches of Demosthenes and especially in *De Corona*, a number of such documents are present, professing to be genuine legal documents of the fourth century B.C. Among these documents are several psephismata which date a meeting of a legislative assembly to a certain day (e.g., [Dem.] 18.181, which dates a meeting of the Ekklesia to Skirophorion 16). If these documents were genuine, they would provide a valuable source of data for this calendric study.

Unfortunately, however, these documents are forgeries, and have been widely recognized as such for over 100 years. A. Westermann in 1850 (*Abhandlung der sächsischen Gesellschaft der Wissenschaften*, Leipzig, 1850, I, pp. 1–136) initiated the critical study of these documents, and successfully demonstrated that they are forgeries. This fact is so widely accepted today that modern editors of Demosthenes without comment denote them as forgeries.

The reasons for denoting them as forgeries are numerous, and more than sufficient. The psephismata are not written according to the rigid formulae of genuine psephismata as we know them from inscriptions. The archons listed in these psephismata cannot be dated to the purported year of the psephismata, and/or are nowhere else recorded as having been archons. The content of the psephismata is either completely incorrect, or has been drawn from the speech itself.

The pseudo-psephismata which relate to this study have been included in this appendix for purposes of reference. The widespread acceptance of them as forgeries allows me to pass over them with only the briefest comments.

Hekatombaion 3

[Demosthenes] 18.137 would suggest legal proceedings on this day, but the first archon named Nikias is dated to 296/5 B.C.

APPENDIX II

Hekatombaion 30

[Demosthenes] 18.29 records this day as a meeting day of the Ekklesia in 347/6 B.C. W. W. Goodwin (*Demosthenes: On the Crown*, Cambridge, England, 1901, pp. 29–30, note 29) demonstrates that this particular psephisma is a "good specimen of ignorant forgery."

Boedromion 16

[Demosthenes] 18.105 assigns a meeting of the Ekklesia to this day. The archon Polykles named there is unknown.

Boedromion 25

[Demosthenes] 18.115 dates a meeting of the Ekklesia to this day. The archon Demonikos named there is unknown.

Boedromion 30

[Demosthenes] 18.75 dates a meeting of the Boule to this day. The archon Neokles named there is unknown.

Pyanopsion 22

[Demosthenes] 18.118 dates a meeting of the Ekklesia to this day. Euthykles, the archon named there, cannot be dated to the period of the psephisma.

Maimakterion 21

[Demosthenes] 18.37 dates a meeting of the Boule to this day. The archon Mnesiphilos named there is unknown.

Gamelion 25

[Demosthenes] 18.84 dates a meeting of the Ekklesia to this day.

Elaphebolion 6

[Demosthenes] 18.54 would indicate that legal proceedings occurred on this day in 338/7 B.C., but Goodwin (*Demosthenes: On the Crown*, p. 43) has demonstrated that this γραφή is a forgery.

PSEUDO-PSEPHISMATA

Elaphebolion 25

[Demosthenes] 18.164 dates a meeting of the Boule to this day. Heropythes, the archon named there, is unknown.

Skirophorion 16

[Demosthenes] 18.181 dates a meeting of the Ekklesia to this day. Nausikles, the archon named there, is unknown.

INDEX OF DEITIES, HEROES, AND FESTIVALS

Agathos Daimon, Day 2 of each month, 15, 24, 186, 190-192, 196-197, 214
Aglauros, 166-167
Akheloos, 63
Alokhos, 63
Amazons, 71
Anakes, 152
Anarrusis, 79
Anemoi, 93
Anodos, 71-73, 187, 189
Anthesteria, 3, 113-114, 185, 187, 190
Apatouria, 6, 79
Aphrodite, Day 4 of each month, 16-18, 24, 186, 191-192, 196-197, 214
Apollo, Day 7 of each month, 13-14, 16, 19-20, 24, 140, 186, 190, 214
―――― Apotropaios, 20, 26, 99-100
―――― Boedromios, 51
―――― Delphinios, 19, 98-99
―――― Hekatombaios, 26
―――― Lykeios, 19, 38, 98-99
―――― Metageitnios, 36
―――― Nymphegetes, 20, 99-100
―――― Paion, 152
―――― Patroos, 19, 36
―――― Pythios, 69, 152
Areopagos, 22-23, 62
Arrephoria, 167
Artemis, Day 6 of each month, 18-19, 21, 24, 32, 36, 70, 143-144, 186, 191-192, 197
―――― Agrotera, 18, 50, 186, 197
―――― Hekate, 40
―――― Mounichia, 144
―――― Soteira, 140
Asklepieia, 6, 46, 70
Asklepios, 4, 20, 58, 123, 187, 189, 191
Athena, Day 3 of each month, 16, 19, 23-24, 36, 47, 63, 79, 101, 160, 163, 186, 191-192, 197, 214

―――― Phratria, 30
―――― Polias, 38-39, 166-167
―――― Skiras, 79, 170

Basile, 49
Bendideia, 26, 145, 158, 160, 163, 180, 187, 192
Bendis, 111, 165, 168-169
Boedromia, 19, 51
Bouphonia, 171

Chalkeia, 78, 187, 191-192
Charites, 16

Dekadistai, 14
Delphinia, 30, 140, 150, 186, 214
Demeter, 38, 56
Demokratia, 46, 53, 187, 192
Diasia, 117, 120, 187
Diisoteria, 180
Dionysia
―――― in the city, 5-6, 109-110, 123-130, 137, 145, 185, 187, 189-192
―――― in the Piraeus, 6, 70, 97
―――― rural, 88, 93, 97, 109
Dionysos, 15, 39-40, 58, 104, 109-111, 113, 124-126, 129-130
―――― Leneus, 110
Dipolieia, 171, 187, 191
Dorpeia, 79

Eikadistai, 14
Eirene, 31
Eleusinia (festival), 6, 40, 46, 199
Eleusinia (goddess), 46
Epibda, 79
Epidauria, 56-58, 65
Epimenia, 15
Epops, 49-50
Eranistai, 144-145

219

INDEX

Eros, Day 4 of each month, 16-18, 24, 186, 191-192, 196-197, 214
Eurysakes, 145

Gamelia, 107
Ge, 63, 126-127
Genesia, 49-50, 187, 192

Haloa, 3, 94-95, 187, 192
Hekate, 23-24, 40
Hekatombaia, 19, 26
ἡμέραι ἀποφράδες, 5, 22-23
ἡμέραι ἀφέσιμοι, 193-197
Hera, 107, 189
——— Thelkhinia, 42
Herakleidai, 17, 139
Herakles, Day 4 of each month, 16-17, 24, 149, 186, 191-192, 196-197, 214
Hermaphrodites, 18
Hermes, Day 4 of each month, 16-18, 20, 24, 63, 186, 191-192, 196-197, 214
——— Hegemonios, 145
Heroines, 41, 73

Iakkhos, 55, 59
Iobakkhoi, 123-124, 126

Kalligeneia, 71-73, 187
Kallynteria, 160, 163-164
Karneia, 153
Kathodos, 71-73
Khoes, 113-114
Khytroi, 113-114
Konnidas, 70
Kore, 46, 56
Koureotis, 79
Kourotrophos, 32, 40, 98-99, 107, 166-167
Kronia, 4, 28, 187, 189, 203
Kronos, 129
Kybernesia, 6, 71

Lenaia, 6, 109-110
Leto, 19, 36, 152
Leukaspis, 146

Maia, 17
Marathonia, 6
Menedeios, 158
Metageitnia, 36
Metoikia, 30
Mounichia, 6, 21, 143-144, 150, 187
Mysteries
——— at Agrai, 6, 120-121
——— at Eleusis, 21, 54-62, 65, 187, 192

Nephthys, 54
Nesteia, 71-74, 187
Niketeria, 47, 186, 212
Noumenia, Day 1 of each month, 8-9, 14-16, 19-20, 24, 186, 190-191, 214
Noumeniastai, 14
Nymphs, 20, 63, 99-100

Olympieia, 6, 145, 150, 187
Osiris, 54
Oskhophoria, 30, 68-69

Panathenaia, 4, 16, 22-23, 28, 30, 34, 46, 185, 187, 199, 203
Pandia, 137
Pandrosos, 167
Pithoigia, 113
Plemokhoai, 65
Plerosia, 88
Plynteria, 6, 22, 160, 163-164, 187
Pompaia, 86
Poseidon, Day 8 of each month, 15-16, 19-20, 24, 47, 107, 166-167, 186, 191-192, 197, 214
Posidea, 89
Proagon, 4, 6, 123, 125, 187, 189, 191
Proerosia, 67-69
Pyanopsia, 19, 30, 68-70, 186

Semele, 129-130
Skira, 71, 170, 187, 191
Soteriastai, 18, 140
Stenia, 71, 187
Synoikia, 29-31, 187, 190, 203

220

INDEX

Tetradistai, 17
Thargelia, 19, 153-154, 186
Theogamia, 22, 107, 187, 189
Theseia, 69-71, 114, 186
Theseus, Day 8 of each month, 16, 20, 24, 30, 68-71, 79, 186, 191-192, 197, 214
Thesmophoria, 6, 71-74, 187, 189
Tritopatores, 147

Zeus, 15, 63, 88, 91-92, 145, 152
——— Eleutherios, 48
——— Epakrios, 157
——— Epopetes, 44
——— Georgos, 6
——— Horios, 91-92
——— Meilikhios, 117, 120
——— Phratrios, 30, 79
——— Polieus, 38-39, 166-167, 171
——— Soter, 177, 180
——— Teleios, 107

INDEX OF INSCRIPTIONS

American Journal of Philology
63 (1942), p. 422: 51, 112, 192

Bulletin de Correspondance Hellénique
59 (1935), pp. 64-70: 147
87 (1963), pp. 603-634: 32, 38-42, 44, 49-50, 63, 73, 86, 91-92, 98-101, 107, 111, 117, 129-130, 139, 146-147, 152, 157-158, 166-167

Dow, *Prytaneis*
pp. 38-39, no. 4: 84
pp. 86-88, no. 38: 147
pp. 91-92, no. 41: 118-119, 127, 192
pp. 109-110, no. 53: 122
pp. 112-113, no. 56: 105, 108
pp. 120-124, no. 64: 76, 81
pp. 133-135, no. 72: 96
pp. 142-146, no. 79: 105, 111

Hesperia
1932, pp. 43-44: 138-139
1932, pp. 45-56: 45
1933, pp. 156-158, no. 5: 179
1933, pp. 160-161, no. 7: 52-53
1933, pp. 163-165, no. 9: 39, 41
1933, pp. 398-402, no. 18: 65
1934, pp. 3-4, no. 5: 148
1934, pp. 6-7, no. 7: 95
1934, pp. 14-18, no. 17: 72, 78, 124-125, 192
1934, pp. 18-21, no. 18: 180
1934, pp. 21-27, no. 19: 80
1934, pp. 27-31, no. 20: 115
1934, pp. 31-35, no. 21: 36
1935, p. 21, no. 2: 29-30, 162, 165
1935, pp. 35-37, no. 5: 109
1935, pp. 37-38, no. 6: 141
1935, pp. 71-81, no. 37: 50, 52, 62, 133, 139, 192
1935, pp. 525-530, no. 39: 45
1935, pp. 562-565, no. 40: 45

1936, pp. 201-203: 158
1936, pp. 393-413, no. 10: 66, 165
1936, pp. 413-414, no. 11: 90
1936, pp. 418-419, no. 14: 57
1936, pp. 419-428, no. 15: 128
1938, pp. 3-5: 36, 68-69, 79, 145
1938, pp. 97-100, no. 17: 150
1938, pp. 100-109, no. 18: 100
1938, pp. 291-292, no. 18: 166, 190
1938, pp. 292-294, no. 19: 172
1938, pp. 114-115, no. 21: 53, 74, 91, 143, 192
1938, p. 297, no. 22: 178
1938, pp. 121-123, no. 24: 57
1938, pp. 123-126, no. 25: 134
1938, pp. 476-479, no. 31: 127-128
1939, pp. 26-27, no. 6: 131
1939, pp. 30-34, no. 8: 103-104, 109
1939, p. 42, no. 10: 116
1940, pp. 80-83, no. 13: 126, 134
1940, p. 83, no. 14: 52
1940, pp. 84-85, no. 15: 181
1940, pp. 85-86, no. 16: 172
1940, pp. 104-111, no. 20: 77
1940, pp. 325-327, no. 35: 99
1940, pp. 327-328, no. 36: 90
1940, pp. 345-348, no. 44: 81-82
1941, pp. 14-27, no. 1: 210
1941, pp. 49-50, no. 12: 95
1941, pp. 268-270, no. 69: 181
1941, pp. 275-277, no. 73: 160
1941, pp. 338-339: 131
1942, pp. 282-287, no. 55: 31
1942, pp. 293-298, no. 58: 66
1944, pp. 234-241, no. 6: 10, 98, 101, 140, 191
1946, pp. 140-142, no. 3: 80
1946, pp. 144-146, no. 6: 116, 173
1946, pp. 206-211: 97
1947, p. 163, no. 61: 142
1947, pp. 170-172, no. 67: 75
1947, p. 187, no. 93: 176
1948, pp. 1-2, no. 1: 122, 128, 191-192

INDEX

1948, pp. 3-4, no. 3: 124
1948, pp. 17-22, no. 9: 85
1948, pp. 142-143: 27
1952, pp. 367-368, no. 8: 181
1954, pp. 287-296: 37
1957, pp. 33-47, no. 6: 43
1957, pp. 54-55, no. 11: 169-170
1957, pp. 68-71, no. 20: 43
1957, pp. 72-77, no. 22: 131
1960, p. 76, no. 153: 41
1960, pp. 76-77, no. 154: 126
1961, p. 229, no. 28: 173
1961, pp. 229-230, no. 29: 173
1961, pp. 289-292, no. 84: 11, 85-86
1963, pp. 5-6, no. 6: 95, 100, 191
1963, pp. 22-23, no. 23: 81, 192
1963, pp. 23-24, no. 24: 60
1964, pp. 170-171, no. 25: 126
1964, pp. 183-184, no. 34: 155
1964, pp. 200-201, no. 52: 148
1969, pp. 418-425, no. 1: 27
1969, pp. 425-431, no. 2: 82-83, 88

Supplement IV, pp. 144-147: 106

Inschriften von Magnesia am Maeander

no. 37: 67

Inscriptiones Graecae I²

6: 120
298: 32
304B: 209-214
305: 213
324: 33
328: 211
842: 153

Inscriptiones Graecae II²

204: 91
229: 181
237: 161-162
239: 99, 101, 190
242: 181
276: 181
328: 85

330: 64, 143, 181
331: 64, 181
332: 133
333: 168
335: 62, 148
336: 82
336b: 136
338: 37
339a: 40, 44
340: 82
343: 94
344: 52
345: 131
346: 131
347: 131
348: 131
349: 169
350: 112-113, 117-118
351: 151, 155, 190-191
352: 156
353: 6, 78, 191
354: 136
356: 85, 119
357: 96, 163
359: 124, 191
361: 159
362: 156, 158
363: 113, 115, 190
365: 27
367: 75-77
368: 52, 90, 96
372: 129, 131
373: 151
375: 162-163, 181
378: 88, 94, 191
380: 53
383b: 64, 162
388: 137, 142-143
389: 147, 150
390: 171, 191
405: 148
414a: 148
415: 178
420: 31, 190
448: 83-84, 86, 92, 94, 96
450: 6, 102

INDEX

452: 103-104
453: 104
454: 176-177
455: 37, 151-152, 168, 191
456: 86
459: 116
460: 126, 151-152, 167, 191
461: 126, 131-132, 135
462: 126
470: 109
472: 150
477: 87, 192
481: 75
482: 95
483: 108
484: 113
485: 6, 161
486: 178
489: 112, 191
490: 115
493: 174
494: 174
495: 178
496: 178
497: 178
498: 168, 191
499: 102
500: 119
501: 119
502: 148-149
503: 157
504: 177
505: 174
547: 78, 162
562: 119
585: 162
597: 181
640: 43
641: 42
642: 159
644: 6, 144
646: 6, 126
647: 124
649: 149-150
650: 27
651: 116

653: 109
654: 176
655: 176
656: 122
657: 6, 57
659: 178
660: 163
661: 120
662: 136
664: 169, 192
665: 61
666: 6, 90
669: 83-84
670: 103, 127, 170, 191
672: 146
674: 78
675: 114
676: 174
679: 74
680: 133
684: 44, 64
685: 179
687: 37
689: 90, 102
697: 133, 174
700: 64
702: 84
703: 136
704: 135
734: 41
766: 53, 74, 91, 143
768: 6, 142
769: 12, 77
770: 155
772: 175
774: 170
775: 6, 146
777: 143
778: 38
780: 132-133
781: 132
782: 104
783: 160, 176
784: 169
787: 6, 57-58
788: 115

INDEX

790: 156
795: 74
796: 11, 42
797: 65, 120
799: 59, 65, 192
832: 120
833: 159
837: 61
839: 83, 86
843: 154
847: 46, 166
848: 60-64
849: 106-107, 189
850: 136
852: 58
857: 132
864: 43
886: 6, 102
888: 116
889: 175
890: 94
891: 41-42, 142, 146
892: 149
893: 172
896: 132-133
897: 141
898: 142
905: 6, 141-142
910: 106
911: 181
912: 48
915: 35
916: 177, 179
917: 87
945: 180
947: 142
949: 172
950: 172
951: 95, 161
952: 167
953: 96
954: 149
955: 149
967: 133-134
968: 6
971: 175

973: 38
974: 108, 161
977: 105
978: 6, 116
989: 90
991: 133
996: 142
1003: 95, 192
1006: 6, 72, 189
1008: 6, 52, 125
1009: 6, 89
1011: 6, 54-55, 74, 102, 157
1012: 6, 99
1014: 6, 67
1019: 37
1027: 53, 72
1028: 46, 50-52
1034: 102
1036: 38
1039: 6, 50-51
1040: 64
1043: 48
1046: 175
1072: 60
1077: 96
1078: 54, 58-59
1183: 88, 93, 97
1227: 44-45
1263: 66
1277: 144
1282: 33
1283: 26
1284: 168-169
1317: 165
1317b: 165
1330: 39-40
1343: 140
1357a: 49
1358: 26, 46, 126-127
1363: 67-70
1367: 20, 40, 54, 56, 58, 70, 89, 93, 104, 129, 149
1368: 123-124, 126
1369: 144-145
1492: 210, 212-214
1496: 31, 46, 53, 70, 145, 180

INDEX

1578: 29
1589: 210
1629: 138-139
1672: 25, 95, 203
1673: 159
1678: 48
1682: 211
2336: 46
2501: 55
3079: 145

Inscriptiones Graecae IV²

83: 62
84: 58

Inscriptiones Graecae VII

4252: 154-155
4253: 154-155
4254: 74

Inscriptions de Délos

1497: 106
1497 *bis*: 122
1498: 101
1501: 93

1502: 112
1503: 143
1504: 92-93
1505: 162
1506: 171-172
1507: 119

Meritt, *Year*

pp. 192-194: 179
pp. 194-195: 174

Pritchett and Meritt, *Chronology*

pp. 7-8: 28
pp. 23-27: 126
pp. 100-101: 177
pp. 110-111: 65
pp. 117-118: 141
pp. 121-123: 134
pp. 123-126: 159, 192

Supplementum Epigraphicum Graecum

2, no. 9: 111
2, no. 10: 165
21, no. 272: 173

Library of Congress Cataloging in Publication Data

Mikalson, Jon D 1943–
    The sacred and civil calendar of the Athenian year.

    1. Calendar, Greek. I. Title.
CE42.M52   529′.32′208   74-25622
ISBN 0-691-03458-8